T0225040

Optimization Based Model Using Fuzzy and Other Statistical Techniques Towards Environmental Sustainability

Samsul Ariffin Abdul Karim ·
Evizal Abdul Kadir · Arbi Haza Nasution
Editors

Optimization Based Model Using Fuzzy and Other Statistical Techniques Towards Environmental Sustainability

 Springer

Editors
Samsul Ariffin Abdul Karim
Department of Fundamental
and Applied Sciences
Universiti Teknologi PETRONAS
Seri Iskandar, Perak, Malaysia

Evizal Abdul Kadir
LPPM
Universitas Islam Riau
Pekanbaru, Riau, Indonesia

Arbi Haza Nasution
LPPM
Universitas Islam Riau
Pekanbaru, Riau, Indonesia

ISBN 978-981-15-2654-1   ISBN 978-981-15-2655-8   (eBook)
https://doi.org/10.1007/978-981-15-2655-8

This Springer imprint is published by the registered company Springer Nature Singapore Pte Ltd.
The registered company address is: 152 Beach Road, #21-01/04 Gateway East, Singapore 189721, Singapore

Preface

Universiti Teknologi PETRONAS (UTP), Malaysia and Universitas Islam Riau (UIR), Indonesia, started a joint research program through International Collaborative Research Funding (ICRF) in 2019. This book includes five chapters that summarize some researches of that program which are related to the implementation of information technology and mathematical modelling in solving real-life problems faced by both countries. The results of these researches will benefit both universities and both countries.

Chapter 1 discusses a new fuzzy weighted multivariate regression to predict water quality index at Perak rivers in Malaysia. The model is in line with the National Water Quality Standards for Malaysia. It also has great potential to be implemented in Indonesia. Chapter 2 is also trying to solve the water pollution problem. While the first chapter focuses more on the mathematical prediction model, however, the following chapter focuses more on the infrastructure, i.e. a smart sensor node of wireless sensor networks (WSNs) for remote river water pollution monitoring system.

Moving to Chap. 3, the common problems that want to be solved are about the lack of language resources of both countries. There are 706 living languages in Indonesia and 136 living languages in Malaysia. Most of them are low-resource languages with a potential to be endangered in few years ahead if there is no action taken regarding the preservation, enrichment or creation of the language resources.

Chapter 4 presents the current studies on biomass activated carbon from oil palm shell as a potential material to control filtration loss in water-based drilling fluid. The result indicates that activated carbon is a potential material to control filtration loss in the water-based drilling fluid.

The last chapter explained the experimental study on the performance of a PV/T air solar collector. The operation of the solar collector is observed and studied under the change of parameters which are temperature, solar irradiance and air mass flow rate. The predictive model can help both countries to implement a more efficient solar panel to produce renewable energy for housing usage.

We hope this book will strongly support and encourage researchers who want to implement information technology and mathematical modelling in solving real-life problems. The editors would like to express their gratitude to all the contributing authors for their great effort and full dedication in preparing the manuscripts for the book. Each chapter has been reviewed up to eight times by the reviewers and the editors. This is a very tedious process. Therefore, we would like to thank all reviewers for reviewing all manuscripts and provide very constructive feedback within the given timeframe.

This book is suitable for all postgraduate students and researchers who are working in this rapid growing area. Any constructive feedback can be directed to the first editor. This book is fully supported by Universiti Teknologi PETRONAS (UTP) and Universitas Islam Riau (UIR) through research grant: International Collaboration Research Funding (ICRF). We extend our gratitude to both universities.

Seri Iskandar, Malaysia   Samsul Ariffin Abdul Karim
Riau, Indonesia   Evizal Abdul Kadir
Riau, Indonesia   Arbi Haza Nasution

Contents

Contributors

Mohd Nazari Abu Bakar Faculty of Applied Sciences, Universiti Teknologi MARA, Arau Campus, Arau, Perlis, Malaysia

Nur Hadziqoh Muhammad Amin Faculty of Engineering, Universitas Islam Riau, Pekanbaru, Riau, Indonesia

Arif Rahmadani Amru Faculty of Engineering, Universitas Islam Riau, Pekanbaru, Riau, Indonesia

Beh Hoe Guan Fundamental and Applied Sciences Department, Universiti Teknologi PETRONAS, Seri Iskandar, Malaysia

Toru Ishida School of Creative Science and Engineering, Waseda University, Tokyo, Japan

Evizal Abdul Kadir Department of Informatics, Faculty of Engineering, Universitas Islam Riau, Pekanbaru, Riau, Indonesia

Samsul Ariffin Abdul Karim Fundamental and Applied Sciences Department and Centre for Smart Grid Energy Research (CSMER), Institute of Autonomous System, Universiti Teknologi PETRONAS, Bandar Seri Iskandar, Perak Darul Ridzuan, Malaysia

Noran Nur Wahida Khalili Fundamental and Applied Sciences Department, Universiti Teknologi PETRONAS, Seri Iskandar, Malaysia

Yohei Murakami Faculty of Information Science and Engineering, Ritsumeikan University, Kyoto, Japan

Arbi Haza Nasution Department of Informatics Engineering, Faculty of Engineering, Universitas Islam Riau, Pekanbaru, Indonesia

Mahmod Othman Fundamental and Applied Sciences Department, Universiti Teknologi PETRONAS, Seri Iskandar, Malaysia

Bahruddin Saad Fundamental and Applied Sciences Department, Universiti Teknologi PETRONAS, Seri Iskandar, Malaysia

Hamzah Sakidin Fundamental and Applied Sciences Department, Universiti Teknologi PETRONAS, Seri Iskandar, Malaysia

Abdul Syukur Department of Informatics, Faculty of Engineering, Universitas Islam Riau, Pekanbaru, Riau, Indonesia

Mursyidah Umar Faculty of Engineering, Universitas Islam Riau, Pekanbaru, Riau, Indonesia

Muhammad Irfan Yasin Universiti Teknologi MARA (Terengganu), Kuala Terengganu, Terengganu Darul Iman, Malaysia

Hasnah M. Zaid Fundamental and Applied Sciences Department, Universiti Teknologi PETRONAS, Seri Iskandar, Malaysia

Chapter 1
A New Fuzzy Weighted Multivariate Regression to Predict Water Quality Index at Perak Rivers

Muhammad Irfan Yasin and Samsul Ariffin Abdul Karim

This study proposes a new fuzzy weighted multivariate regression analysis to assess the water pollution based on water quality index (WQI). The data used to test the performance of the proposed approach were collected from several locations of Perak River at Perak State, Malaysia. The water quality index was monitored between 2013 and 2017. WQI is measured by integrating all six (6) parameters such as chemical oxygen demand, ammoniacal nitrogen, dissolved oxygen, suspended solid, pH and biochemical oxygen demand (BOD). A fuzzy approach has shown to be practical, simple and useful tool to assess the water pollution levels. Furthermore, the constructed WQI model is line with the National Water Quality Standards for Malaysia.

1.1 Introduction

Water quality is a basic tool for a good health. It is an important because water resources are the major environmental, social and economic. Poor water quality may cause its value and not only effects the aquatic life but also the surrounding ecosystem. The most common sources of water for irrigation include rivers, reservoirs and lakes, and groundwater. Urban environment can sometimes lead to cause of river pollution.

M. I. Yasin
Universiti Teknologi MARA (Terengganu), Kampus Kuala Terengganu, 21080 Kuala Terengganu, Terengganu Darul Iman, Malaysia

S. A. A. Karim (✉)
Fundamental and Applied Sciences Department and Centre for Smart Grid Energy Research (CSMER), Institute of Autonomous System, Universiti Teknologi PETRONAS, 32610 Bandar Seri Iskandar, Perak Darul Ridzuan, Malaysia
e-mail: samsul_ariffin@utp.edu.my

Various techniques have been applied to determine the water quality index (WQI). Horton proposed the first formal water quality index (WQI) which is to study general indices, selecting and weighting parameters [9]. Delphi technique as a tool in a formal assessment procedure, it is most popular assessment methodology which is developed by The National Sanitation Foundation (NSF) [13]. According to [15], the uncertainties involved in using fuzzy membership with values ranging from 0 to 1 to form an applicable fuzzy set instead of the 0–100 scale used in conventional rating curves in WQI [15].

Fuzzy linear regression is suitable for coping with such fuzzy data or linguistic variables. It was stemming from Zadeh [17] thought that fuzzy that was able to deal with ambiguity. Fuzzy linear regression could be applied to the development of environmental indices in a manner that solves several environmental problems, including the incompatibility of observations, uncertainty, imprecision in criteria and the need for implicitly of value judgments. The fuzzy linear regression method also has been used for many years in developing predictive models for various applications including marketing, management and sales forecasting [8], for example [1] has investigated the relationship between variables impacting car sales volume.

There were many improvements of the fuzzy linear regression method and its applications as well. The most of the existing research on the fuzzy regression model have used the least squares method to construct the fuzzy regression model. However, the least squares method is so sensitive to outliers. Basically, the fuzzy regression models can be classified into two classes. The first class is based on the possibility concept and the second class is the least-squares approach [6, 7].

The regression model proposed by Tanaka et al. [16] is quite popular and useful but this model is restricted to symmetric triangular fuzzy numbers. To overcome this limitation, Chang and Lee developed a fuzzy least-squares regression model but for their studies, the regression coefficient is derived from a nonlinear programming problem that requires considerable computations [5]. Therefore, Pan et al. [14] have introduced a matrix-driven multivariate fuzzy linear regression as part of their efforts to improve computational efficiency. The method has been successfully tested in engineering study of estimating bridge performance. WQI was initially designed to include nine parameters designed for making an integrated assessment of water quality conditions in order to meet utilisation goals. There are including dissolved oxygen, faecal coliforms, pH, biochemical oxygen demand (BOD), nitrates, phosphates, temperature, turbidity, and total solids [4]. Based on the works of [3, 18], as well as the standard prepared by DOE, six parameters as mentioned above are enough to represents the WQI model. Thus, in this study, we will be investigating the effect of all six parameters toward the developing of new fuzzy multivariate regression analysis to determine the relationship between parameters. The parameters are pH, chemical oxygen demand, ammoniacal nitrogen, dissolved oxygen, suspended solid and biochemical oxygen demand (BOD).

The main objective of the present studies is to develop new decision models for Water Quality Index summarized below:

(a) To propose new fuzzy regression model
(b) To find the left and right fuzzy regression equations
(c) To predict river Water Quality Index by using the proposed fuzzy regression model for other rivers and stations.

This chapter is organized as follows. In Sect. 1.1, we give some basic introduction as well as related literature review. Section 1.2 is devoted to the construction of general fuzzy regression model by incorporating the crisp and spreading factor. Section 1.3, discuss the data collection and the tools that have been used to collect the data. Section 1.4 is dedicated for Results and Discussion including the comparison with established regression methods. Section 1.5 explain the prediction model and Summary will be given in the final section.

1.2   General Fuzzy Regression

Figure 1.1 shows the framework of the research methodology that is used in this study.

Fig. 1.1 General data fitting flow chart

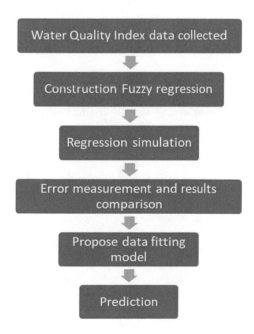

1.2.1 Method and Calculation

The general form of the model can be written as in (1.1).

$$\tilde{y}_i = f(X, A) = \tilde{A}_0 + \tilde{A}_0 X_{ik}^2 + \cdots + \tilde{A}_0 X_{ik}^n \tag{1.1}$$

$\tilde{y}_i$ is the estimated fuzzy response variable or non-fuzzy output data. The function of $f(X, A)$ is consider which mapped from X into Y with the elements of X denoted by $x_i^{(t)} = \left(x_{i0}^t, x_{i1}^t, \ldots, x_{ik}^t \right)$ where $t = 1, 2, \ldots, n$ and ith which represents the independent or input variables of the model. The dependent variables or response variables are denoted as Y_i in, Y where $\tilde{A} = \left(\tilde{A}_0, \tilde{A}_1, \ldots, \tilde{A}_n \right)$ are the model regression coefficients. If $\tilde{A}_j (j = 0, 1, \ldots, n)$ are given as fuzzy sets, the model $f(X, A)$ is called a fuzzy model. $\tilde{A}_j$ is fuzzy coefficient in terms of symmetric fuzzy numbers.

The fuzzy coefficient of triangular fuzzy number (TFN) are shows in Fig. 1.2 that involve its centre or model value, left and right spreads. Asymmetrical TFN that represented by observed data is defined by a triplet $\tilde{y}_i = \left(a_j, c_j^L, c_j^R \right)$.

The data was a symmetrical TFNs, thus $c_j^L = c_j^R = c_j$, Fig. 1.2 also shows the membership function for the symmetrical fuzzy regression coefficients, $A_j = \left(a_j, c_j \right)$. The response variable for symmetric represented $\tilde{y}_i = \left(a_j, c_j \right)$, whereas a_j is the centre point that represents the original value of data and c_j represents for the spread.

The value of the spread, c_j is obtained from [14];

$$c_{n,j} = \frac{|y_i - \bar{y}|}{2} \tag{1.2}$$

This method is considered symmetrical triangular membership function in the following discussion. With k crisp independent variables and one fuzzy dependent variable, the estimated fuzzy quadratic regression can be expressed as

$$\tilde{y}_i = \left(a_0, c_{0,j} \right) + \left(a_1, c_{1,j} \right) X_1 + \left(a_2, c_{2,j} \right) X_2^2 + \cdots + \left(a_k, c_{k,j} \right) X_k^n \tag{1.3}$$

Fig. 1.2 Symmetrical fuzzy regression coefficient

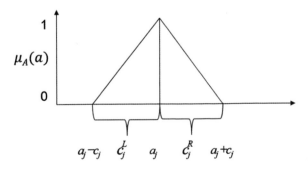

where $(a_0, c_{0,j})$ is the fuzzy intercept coefficient; $(a_1, c_{1,j})$ is the fuzzy slope coefficient for X_1; $(a_2, c_{2,j})$ is the fuzzy slope coefficient for X_2^2; $(a_k, c_{k,j})$ is the kth fuzzy slope coefficient.

The expected $\tilde{y}_i$ at a particular μ value is given by Tanaka et al. [16];

$$
\begin{aligned}
\mu_{\tilde{y},L} &= \left[a_0 - (1 - \mu)c_{0,L}\right] + \left[a_1 - (1 - \mu)c_{1,L}\right]X_1 \\
&\quad + \cdots + \left[a_k - (1 - \mu)c_{k,L}\right]X_k^n \\
&= (a_0 + a_1 + \ldots + a_k) - (1 - \mu)\left(c_{0,L} + c_{1,L}X_1 + \cdots + c_{k,L}X_k^n\right) \quad (1.4)
\end{aligned}
$$

and

$$
\mu_{\tilde{y},R} = (a_0 + a_1 + \cdots + a_k) - (1 - \mu)\left(c_{0,R} + c_{1,R}X_1 + \cdots + c_{k,R}X_k^n\right) \quad (1.5)
$$

in which $a_0, a_1, \ldots, a_k$ are the estimated coefficients of $\tilde{y}_i$ at $\mu = 1$; $c_{0,L} + c_{1,L}X_1$ and $c_{0,R} + c_{1,R}X_1$ are the left fuzzy width and the right fuzzy width for X_1; $c_{0,L} + c_{2,L}X_2^2$ and $c_{0,R} + c_{2,R}X_2^2$ are the left fuzzy width and the right fuzzy width for X_2^2; $(1 - \mu)(c_{0,L} + c_{1,L}X_1)$ and $(1 - \mu)(c_{0,R} + c_{1,R}X_1)$ are the left fuzzy width and the right fuzzy width for X_1 at a given μ value; $(1 - \mu)(c_{0,L} + c_{2,L}X_2^2)$ and $(1 - \mu)(c_{0,R} + c_{2,R}X_2^2)$ are the left fuzzy width and the right fuzzy width for X_2^2 at a given μ value.

The general fuzzy quadratic model can be expressed in the following matrix form:

$$
XB = Y \quad (1.6)
$$

where

$$
X = \begin{bmatrix}
1 & x_{11} & x_{12} & x_{13} & x_{14} & x_{15} & x_{16} \\
1 & x_{21} & x_{22} & x_{23} & x_{24} & x_{25} & x_{26} \\
\vdots & \vdots & \vdots & \vdots & \vdots & \vdots & \vdots \\
1 & x_{n6} & x_{n6} & x_{n6} & x_{n6} & x_{n6} & x_{n6}
\end{bmatrix} \quad (1.7)
$$

$$
B = \begin{bmatrix}
B_0 \\
B_1 \\
B_2 \\
B_3 \\
B_4 \\
B_5 \\
B_6
\end{bmatrix} = \begin{bmatrix}
(a_0, (1 - \mu)c_{0,j}) \\
(a_1, (1 - \mu)c_{1,j}) \\
\vdots \\
(a_n, (1 - \mu)c_{n,j})
\end{bmatrix} \quad (1.8)
$$

and

$$Y = \begin{bmatrix} (y_1, (1-\mu)c_{1,j}) \\ (y_2, (1-\mu)c_{2,j}) \\ \vdots \\ (y_n, (1-\mu)c_{n,j}) \end{bmatrix} \quad (1.9)$$

In the above equations, matrices Y and X are the data matrices associated with response variable and predictor variables, respectively. Matrix B contains the least squares estimates of the regression coefficients. In this study we have adopted a symmetric fuzzy triangular number as discussed in Pan et al. [16] and Isa et al. [10] as in Eqs. (1.7) and (1.9),

To obtain the regression parameters, Eq. (1.1) can be transformed by

$$(X'X)B = X'Y \quad (1.10)$$

where X' is the transpose matrix of X.

The regression coefficients can be derived by matrix operations as follow:

$$B = (X'X)^{-1}X'Y \quad (1.11)$$

where $(X'X)^{-1}$ is the inverse matrix of $X'X$,

$$X'X = \begin{bmatrix} n & \sum x_{1i} & \cdots & \sum x_{6i} \\ \sum x_{1i} & \sum x_{1i}x_{2i} & \cdots & \sum x_{1i}x_{6i} \\ \vdots & \vdots & \ddots & \vdots \\ \sum x_{6i} & \sum x_{1i}x_{6i} & \cdots & \sum x_{6i}^2 \end{bmatrix} \text{ and } Y = \begin{bmatrix} \sum y_i & \sum c_{i,j} \\ \sum x_{1i}y_i & \sum x_{1i}c_{i,j} \\ \vdots & \vdots \\ \sum x_{6i}y_i & \sum x_{6i}c_{i,j} \end{bmatrix}.$$

Besides, the coefficients also can be solved by using Gaussian elimination method.

$$
\begin{bmatrix}
1 & 96.65 & 7 & 22.17 & 0.04 & 8.25 & 6.55 \\
1 & \vdots & \vdots & \vdots & \vdots & \vdots & \vdots \\
1 & \vdots & \vdots & \vdots & \vdots & \vdots & \vdots \\
1 & \vdots & \vdots & \vdots & \vdots & \vdots & \vdots \\
1 & \vdots & \vdots & \vdots & \vdots & \vdots & \vdots \\
1 & \vdots & \vdots & \vdots & \vdots & \vdots & \vdots \\
1 & \vdots & \vdots & \vdots & \vdots & \vdots & \vdots \\
1 & \vdots & \vdots & \vdots & \vdots & \vdots & \vdots \\
1 & \vdots & \vdots & \vdots & \vdots & \vdots & \vdots \\
1 & \vdots & \vdots & \vdots & \vdots & \vdots & \vdots \\
1 & \vdots & \vdots & \vdots & \vdots & \vdots & \vdots \\
1 & \vdots & \vdots & \vdots & \vdots & \vdots & \vdots \\
1 & \vdots & \vdots & \vdots & \vdots & \vdots & \vdots \\
1 & \vdots & \vdots & \vdots & \vdots & \vdots & \vdots \\
1 & 90.22 & 6.5 & 20.6 & 0.26 & 5.5 & 6.39 \\
1 & \vdots & \vdots & \vdots & \vdots & \vdots & \vdots \\
1 & \vdots & \vdots & \vdots & \vdots & \vdots & \vdots \\
1 & \vdots & \vdots & \vdots & \vdots & \vdots & \vdots \\
1 & \vdots & \vdots & \vdots & \vdots & \vdots & \vdots \\
1 & \vdots & \vdots & \vdots & \vdots & \vdots & \vdots \\
1 & \vdots & \vdots & \vdots & \vdots & \vdots & \vdots \\
1 & \vdots & \vdots & \vdots & \vdots & \vdots & \vdots \\
1 & \vdots & \vdots & \vdots & \vdots & \vdots & \vdots \\
1 & \vdots & \vdots & \vdots & \vdots & \vdots & \vdots \\
1 & \vdots & \vdots & \vdots & \vdots & \vdots & \vdots \\
1 & \vdots & \vdots & \vdots & \vdots & \vdots & \vdots \\
1 & 50.48 & 11.5 & 33 & 2.91 & 180.75 & 7.29
\end{bmatrix}
\begin{bmatrix}
B_0 \\ B_1 \\ B_2 \\ B_3 \\ B_4 \\ B_5 \\ B_6
\end{bmatrix}
=
\begin{bmatrix}
87.91 & 4.31 \\
91.63 & 6.17 \\
91.01 & 5.86 \\
89.12 & 4.92 \\
85.25 & 2.98 \\
84.45 & 2.58 \\
80.57 & 0.64 \\
75.94 & 1.67 \\
73.81 & 2.73 \\
75.36 & 1.95 \\
73.46 & 2.90 \\
69.77 & 4.75 \\
75.36 & 1.95 \\
68.72 & 5.27 \\
72.62 & 3.32 \\
85.92 & 3.32 \\
85.67 & 3.19 \\
79.42 & 0.07 \\
74.95 & 2.16 \\
79.04 & 0.11 \\
89.05 & 4.88 \\
93.42 & 7.07 \\
91.92 & 6.32 \\
93.09 & 6.90 \\
90.88 & 5.80 \\
68.22 & 5.52 \\
67.98 & 5.65 \\
77.01 & 1.13 \\
55.03 & 12.12 \\
51.73 & 13.77
\end{bmatrix}
$$

$$
\begin{bmatrix}
1 & 96.65 & 7 & 22.17 & 0.04 & 8.25 & 6.55 \\
1 & \vdots & \vdots & \vdots & \vdots & \vdots & \vdots \\
1 & \vdots & \vdots & \vdots & \vdots & \vdots & \vdots \\
1 & \vdots & \vdots & \vdots & \vdots & \vdots & \vdots \\
1 & \vdots & \vdots & \vdots & \vdots & \vdots & \vdots \\
1 & \vdots & \vdots & \vdots & \vdots & \vdots & \vdots \\
1 & \vdots & \vdots & \vdots & \vdots & \vdots & \vdots \\
1 & \vdots & \vdots & \vdots & \vdots & \vdots & \vdots \\
1 & \vdots & \vdots & \vdots & \vdots & \vdots & \vdots \\
1 & \vdots & \vdots & \vdots & \vdots & \vdots & \vdots \\
1 & \vdots & \vdots & \vdots & \vdots & \vdots & \vdots \\
1 & \vdots & \vdots & \vdots & \vdots & \vdots & \vdots \\
1 & \vdots & \vdots & \vdots & \vdots & \vdots & \vdots \\
1 & \vdots & \vdots & \vdots & \vdots & \vdots & \vdots \\
1 & \vdots & \vdots & \vdots & \vdots & \vdots & \vdots \\
1 & 90.22 & 6.5 & 20.6 & 0.26 & 5.5 & 6.39 \\
1 & \vdots & \vdots & \vdots & \vdots & \vdots & \vdots \\
1 & \vdots & \vdots & \vdots & \vdots & \vdots & \vdots \\
1 & \vdots & \vdots & \vdots & \vdots & \vdots & \vdots \\
1 & \vdots & \vdots & \vdots & \vdots & \vdots & \vdots \\
1 & \vdots & \vdots & \vdots & \vdots & \vdots & \vdots \\
1 & \vdots & \vdots & \vdots & \vdots & \vdots & \vdots \\
1 & \vdots & \vdots & \vdots & \vdots & \vdots & \vdots \\
1 & \vdots & \vdots & \vdots & \vdots & \vdots & \vdots \\
1 & \vdots & \vdots & \vdots & \vdots & \vdots & \vdots \\
1 & \vdots & \vdots & \vdots & \vdots & \vdots & \vdots \\
1 & 50.48 & 11.5 & 33 & 2.91 & 180.75 & 7.29
\end{bmatrix}
\begin{bmatrix}
B_0 \\ B_1 \\ B_2 \\ B_3 \\ B_4 \\ B_5 \\ B_6
\end{bmatrix}
=
\begin{bmatrix}
87.91 & 2 \\
91.63 & 2 \\
91.01 & 2 \\
89.12 & 2 \\
85.25 & 2 \\
84.45 & 2 \\
80.57 & 2 \\
75.94 & 2 \\
73.81 & 2 \\
75.36 & 2 \\
73.46 & 2 \\
69.77 & 2 \\
75.36 & 2 \\
68.72 & 2 \\
72.62 & 2 \\
85.92 & 2 \\
85.67 & 2 \\
79.42 & 2 \\
74.95 & 2 \\
79.04 & 2 \\
89.05 & 2 \\
93.42 & 2 \\
91.92 & 2 \\
93.09 & 2 \\
90.88 & 2 \\
68.22 & 2 \\
67.98 & 2 \\
77.01 & 2 \\
55.03 & 2 \\
51.73 & 2
\end{bmatrix}
$$

1.2.2   Error calculation

To prove the effectiveness of clustering in predicting the quality of water, this experiment took out the highest and lowest data from the test data to run it on the model produced in (1.6). To present the results in a comparative way, the absolute error e_i i.e. error of deviation between theoretical value Y_i and the experimental value $\hat{Y}_i$, is used as the metric in comparing this result. It is defined as general in (1.12),

$$e_i = \frac{\left|Y_i - \hat{Y}_i\right|}{Y_i} \tag{1.12}$$

1.3   Data Collection

The case study is situated in the state of Perak, Malaysia, where there are 11 major river basins that covers over 80 km². The Perak River basin is about 760 km long with an area of 14.908 km². Perak River basin, shown in Figs. 1.3, 1.4 and 1.5, is the biggest river basin in this area, which covers about 70% of the state area.

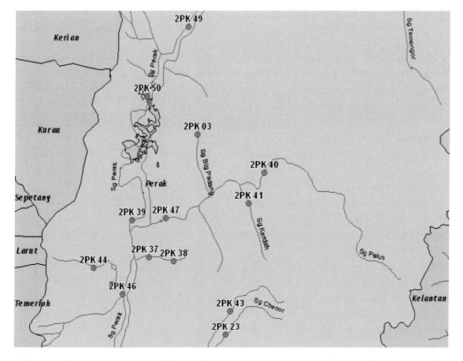

Fig. 1.3 Perak (North) river basin monitoring stations [2]

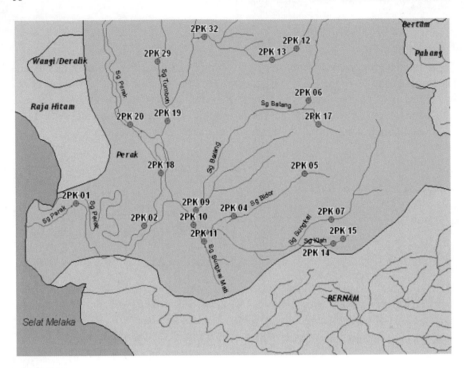

Fig. 1.4 Perak (South) river basin monitoring stations [2]

If water in the Perak River basin is contaminated it will affect most of the river basin in Perak State and affect the human population as well as the financial income of the local population where most of the activity in this area is fishing and agriculture activities. The purpose of the study is to predict the water quality of the Perak River basin in real time. If poor river water quality is predicted, some preventive measures can be taken immediately.

The sample and data collection were duly carried out by the Department of Environment (DOE) of Malaysia through Alam Sekitar Malaysia Sdn. Bhd. [2]. The concerned area in the Perak river basin is divided into 3 sections which are north, south and central area as shown respectively in Figs. 1.3, 1.4 and 1.5 and it covers about 21 rivers along the Perak river basin. The samples are taken at selected point or location using an automatic sampling approach as well as are manually taken once a week or twice a month.

The DOE of Malaysia introduced the WQI monitoring approach in 1978 that considers the six variables, which are dissolved oxygen (DO), biological oxygen demand (BOD), chemical oxygen demand (COD), suspended solid (SS), the pH value (pH), and ammonical nitrogen (NH_3–NL) [11]. DOE applies the following formula for the calculation of WQI [12]:

$$WQI = (0.22 * SIDO) + (0.19 * SIBOD) + (0.16 * SICOD)$$

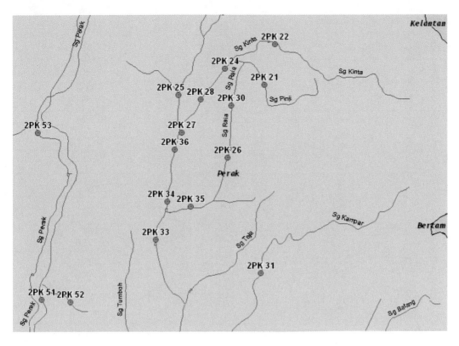

Fig. 1.5 Perak (Central) river basin monitoring stations [2]

$$+ (0.15 * SIAN) + (0.16 * SISS) + (0.12 * SIpH) \qquad (1.13)$$

where;

$$SIDO = Subindex \ DO \ (\% \, saturation)$$
$$SIBOD = Subindex \ BOD$$
$$SICOD = Subindex \ COD$$
$$SIAN = Subindex \ NH_3 - NL$$
$$SISS = Subindex \ SS$$
$$SIpH = Subindex \ pH$$
$$0 \leq WQI \leq 100$$

1.4   Results and Discussion

In this section, the results and findings from the models, techniques, and algorithms developed are presented and discussed. The main emphasis is on the decision models of river WQI. The experiments and testing were conducted using the actual data of

Table 1.1 Fuzzy multivariate regression and spreading value based on formula

Data river	Theoretical value, Y_i	Spreading value, $c_{n,j}$
1	87.9138	4.3161
2	91.6313	6.1749
3	91.0139	5.8662
4	89.1231	4.9208
5	85.2555	2.9870
6	84.4588	2.5886
7	80.5775	0.6480
8	75.9404	1.6704
9	73.8192	2.7311
10	75.3687	1.9563
11	73.4648	2.9082
12	69.7780	4.7517
13	75.3666	1.9574
14	68.7242	5.2786
15	72.6273	3.3270
16	85.9220	3.3202
17	85.6703	3.1944
18	79.4226	0.0705
19	74.9503	2.1655
20	79.0427	0.1193
21	89.0594	4.8889
22	93.4267	7.0726
23	91.9237	6.3211
24	93.0951	6.9068
25	90.8833	5.8009
26	68.2265	5.5274
27	67.9805	5.6504
28	77.0118	1.1348
29	55.0350	12.1232
30	51.7301	13.7756

WQI from every river located in Kampar, Kinta, Kurau, Manjong, Manong, and Nyamok, which have been processed and prepared in the fuzzy regression model.

Table 1.1 shows a snapshot of the data before preprocessing. The spreading value was calculated using formula in (1.2).

Now, the WQI value for some river located in Perak river basin as shown in Fig. 1.5 can be calculated by using Eqs. (1.7)–(1.9). The result for WQI coefficients are shown below:

$$
B = \begin{bmatrix}
11.56 & 0.0151 & -0.1493 & -0.0057 & 0.364 & 0.0087 & -1.766 \\
0.015 & 0.0001 & 0.0006 & -0.0001 & 0 & 3 \times 10^{-5} & -0.004 \\
-0.14 & 0.0006 & 0.08671 & -0.025 & -0.004 & 3 \times 10^{-5} & 0.008 \\
-0.005 & -0.0001 & -0.0255 & 0.0089 & -0.006 & -6 \times 10^{-5} & 0 \\
0.364 & 0.0004 & -0.0041 & -0.006 & 0.120 & -0.0004 & -0.039 \\
0.008 & 3 \times 10^{-5} & 3 \times 10^{-5} & -6 \times 10^{-5} & 0 & 3 \times 10^{-5} & -0.001 \\
-1.76 & -0.0044 & 0.0089 & 0.0002 & -0.039 & -0.0016 & 0.31
\end{bmatrix}^{-1}
$$

$$
\times \begin{bmatrix}
2378.44 & 130.15 \\
189,944.8 & 10,031.1 \\
15,386.83 & 950.23 \\
48,850.6 & 2973.21 \\
951.7812 & 113.40 \\
80,734.04 & 6641.23 \\
16,310 & 901.77
\end{bmatrix}
$$

$$
B = \begin{bmatrix}
70.4154 & 6.0301 \\
0.2662 & 0.0402 \\
-1.1072 & 1.3404 \\
0.0709 & -0.5635 \\
-2.8051 & 3.3812 \\
-0.0535 & 0.0085 \\
-0.3425 & -0.5506
\end{bmatrix}
$$

The result of fuzzy multivariate regression provided the three models that were categorised as Left, Crisp and Right. The proposed generic regression model (WQI) is to predict other extreme river. Table 1.2 shows the Crisp and the spreading value for the Left and Right model.

Fuzzy multivariate regression form for membership function, $\mu = 0$;

$$
\begin{aligned}
Y = {} & (70.4154, 6.0301) + (0.2662, 0.0402)DO + (-1.1072, 1.3404)BOD \\
& + (0.0709, -0.5635)COD + (-2.8051, 3.3812)AN \\
& + (-0.0535, 0.0085)SS + (-0.3425, -0.5506)pH
\end{aligned}
$$

Table 1.2 Fuzzy multivariate regression coefficient

	B_0	B_1	B_2	B_3	B_4	B_5	B_6
a_j	70.4154	0.2662	−1.1072	0.0709	−2.8051	−0.0535	−0.3425
c_j	6.0301	0.0402	1.3404	−0.5635	3.3812	0.0085	−0.5506

Fuzzy multivariate regression can be written as follows:

$$Y_L = 64.3853 + 0.2259DO - 2.4476BOD - 0.4925COD$$
$$- 6.1863AN - 0.062SS - 0.8932pH$$

$$Y_R = 76.4455 + 0.3064DO + 0.2331BOD + 0.6344COD$$
$$+ 0.576AN - 0.0449SS + 0.208pH$$

where;

$$Y_L = \text{Left Fuzzy multivariate Regression}$$
$$Y_R = \text{Right Fuzzy multivariate Regression}$$

and the crisp equation i.e. the standard multiple linear fitting is given as:

$$Y = 70.4154 + 0.2662DO - 1.1072BOD + 0.0709COD$$
$$- 2.8051AN - 0.0535SS - 0.3425pH$$

The result of WQI value provided by all three fuzzy models are shown in Table 1.3. Experiments were performed using the given spreading formula for each data sets. Figure 1.6 showed that the result is not acceptable because the spreading of Left and Right fuzzy models has a large gap compared with the crisp model which have smaller error with the respective WQI value. Thus, we need to improve the model. We propose a new method called as fuzzy convex combination model.

1.4.1 An Enhanced Fuzzy Multivariate Regression for WQI Data Sets

In order to apply the combination of convex combination and WQI fuzzy multivariate regression, we calculate new spreading value based on the possibilities and uncertainties of the data. Table 1.4 shows the value of new spreading value.

The coefficients of new spreading can be calculated as below:

$$B = \begin{bmatrix} 11.56 & 0.0151 & -0.1493 & -0.0057 & 0.3644 & 0.0087 & -1.766 \\ 0.015 & 0.0001 & 0.0006 & -0.0001 & 0.0004 & 3 \times 10^5 & -0.004 \\ -0.14 & 0.0006 & 0.086 & -0.025 & -0.004 & 3 \times 10^5 & 0.008 \\ -0.005 & -0.0001 & -0.025 & 0.008 & -0.006 & -6 \times 10^5 & 0.0002 \\ 0.364 & 0.0004 & -0.004 & -0.006 & 0.1202 & -0.0004 & -0.039 \\ 0.008 & 3 \times 10^5 & 3 \times 10^5 & -6 \times 10^5 & -0.0004 & 3 \times 10^5 & -0.001 \\ -1.76 & -0.004 & 0.008 & 0.0002 & -0.039 & -0.001 & 0.31 \end{bmatrix}^{-1}$$

Table 1.3 Fuzzy multivariate regression form for membership function and its class

River	Y_L	Y	Y_R	WQI
1	51.5369	87.1597	122.7825	87.9138
2	63.0119	89.1762	115.3405	91.6313
3	64.2586	91.5903	118.9220	91.0139
4	57.8242	89.6834	121.5427	89.1231
5	48.7588	86.0918	123.4248	85.2555
6	57.0497	83.0453	109.0409	84.4588
7	52.4091	82.2916	112.1741	80.5775
8	42.6846	74.1293	105.5740	75.9404
9	38.8582	73.0224	107.1865	73.8192
10	41.0821	75.8986	110.7151	75.3687
11	35.8307	74.1855	112.5404	73.4648
12	39.8724	71.5721	103.2718	69.7780
13	42.7412	75.7414	108.7417	75.3666
14	28.0564	67.4502	106.8439	68.7242
15	44.7138	73.8359	102.9579	72.6273
16	51.0521	85.4829	119.9138	85.9220
17	56.9706	86.2252	115.4797	85.6703
18	33.8762	79.5356	125.1949	79.4226
19	27.9370	75.7089	123.4808	74.9503
20	42.8456	80.1872	117.5289	79.0427
21	53.0800	89.0130	124.9459	89.0594
22	67.2187	91.4577	115.6967	93.4267
23	65.5583	93.0699	120.5815	91.9237
24	67.8171	93.2642	118.7113	93.0951
25	61.6627	90.8697	120.0766	90.8833
26	22.6386	67.0572	111.4759	68.2265
27	17.2119	68.0277	118.8434	67.9805
28	43.9186	77.0352	110.1518	77.0118
29	−16.7068	53.5102	123.7273	55.0350
30	−4.3460	53.1251	110.5963	51.7301

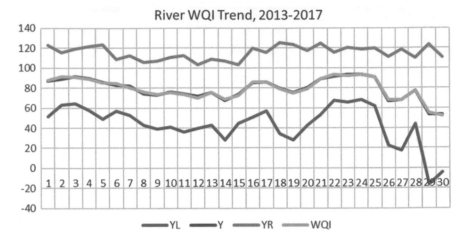

Fig. 1.6 Result for spreading value using given formula of $c_{n,j}$ for membership, $\mu = 0$

$$\times \begin{bmatrix} 2378.44 & 60 \\ 189,944.9 & 4640.69 \\ 153,86.83 & 404.5 \\ 48,850.6 & 1282.31 \\ 951.78 & 29.16 \\ 80,734.05 & 2315.08 \\ 16,310 & 410.92 \end{bmatrix}$$

$$B = \begin{bmatrix} 70.4154 & 2 \\ 0.2662 & 2.89 \times 10^{15} \\ -1.1072 & -8 \times 10^{15} \\ 0.0709 & 5.26 \times 10^{15} \\ -2.8051 & -3.6 \times 10^{15} \\ -0.0535 & 2.22 \times 10^{16} \\ -0.3425 & -1.4 \times 10^{14} \end{bmatrix}$$

Table 1.5 shows the crisp and the spreading value for every coefficient using the data that has been transformed using the membership function (MF) graph in fuzzy logic.

Fuzzy multivariate regression form for membership function, $\mu = 0$

$$Y_{new} = (70.4154, 2) + \left(0.2662, 2.89 \times 10^{-15}\right)DO + \left(-1.1072, -8 \times 10^{-15}\right)BOD$$
$$+ \left(0.0709, 5.26 \times 10^{-15}\right)COD + \left(-2.8051, -3.6 \times 10^{-15}\right)AN$$
$$+ (-0.0535, 2.22 \times 10^{-16})SS + (-0.3425, -1.4 \times 10^{-14})pH \qquad (1.14)$$

Table 1.4 Fuzzy regression and spreading value based on logic

River	Theoretical value	New spreading, C_j
1	87.9138	2
2	91.6313	2
3	91.0139	2
4	89.1231	2
5	85.2555	2
6	84.4588	2
7	80.5775	2
8	75.9404	2
9	73.8192	2
10	75.3687	2
11	73.4648	2
12	69.7780	2
13	75.3666	2
14	68.7242	2
15	72.6273	2
16	85.9220	2
17	85.6703	2
18	79.4226	2
19	74.9503	2
20	79.0427	2
21	89.0594	2
22	93.4267	2
23	91.9237	2
24	93.0951	2
25	90.8833	2
26	68.2265	2
27	67.9805	2
28	77.0118	2
29	55.0350	2
30	51.7301	2

Table 1.5 Fuzzy regression coefficients for new spreading when membership, $\mu = 0$

	B_0	B_1	B_2	B_3	B_4	B_5	B_6
a_j	70.4154	0.2662	−1.1072	0.0709	−2.8051	−0.0535	−0.3425
c_j	2	2.8×10^{-15}	-8×10^{-15}	5.2×10^{-15}	-3.6×10^{-15}	2.22×10^{-16}	-1.4×10^{-14}

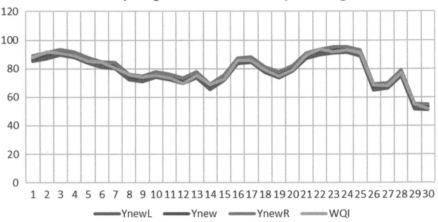

Fig. 1.7 Result for spreading value using new $c_{n,j}$ for membership, $\mu = 0$

where Y_{new} is the Fuzzy multivariate regression model for the modified data set.

Figure 1.7 illustrates the modified data sets that resulting the proposed fuzzy model look pretty good and has shown an improvement compared previous fuzzy model. Table 1.6 shows the value of WQI with respective fuzzy model.

1.5 An Enhanced Fuzzy Convex Combination for Improved Fuzzy Model

The fuzzy convex combination is based on the fuzzy model (1.14) to improve the result for the WQI value. The general form of the model can be written as below,

$$\tilde{Y}_a = (1 - a)Y_{newL} + aY_{newR} \tag{1.15}$$

where Y_{newL} is the fuzzy equation for left spread and Y_{newR} is the fuzzy equation for right spread with the same membership number.

The purpose fuzzy convex combination form for membership function, $\mu = 0$ and $a = 0.4$, and $a = 0.6$,

Table 1.6 Fuzzy multivariate regression form for membership function for $\mu = 0$

River	Y_{newL}	Y_{new}	Y_{newR}	WQI
1	85.1597	87.1597	89.1597	87.9138
2	87.1762	89.1762	91.1762	91.6313
3	89.5903	91.5903	93.5903	91.0139
4	87.6834	89.6834	91.6834	89.1231
5	84.0918	86.0918	88.0918	85.2555
6	81.0453	83.0453	85.0453	84.4588
7	80.2916	82.2916	84.2916	80.5775
8	72.1293	74.1293	76.1293	75.9404
9	71.0224	73.0224	75.0224	73.8192
10	73.8986	75.8986	77.8986	75.3687
11	72.1855	74.1855	76.1855	73.4648
12	69.5721	71.5721	73.5721	69.7780
13	73.7415	75.7415	77.7415	75.3666
14	65.4502	67.4502	69.4502	68.7242
15	71.8359	73.8359	75.8359	72.6273
16	83.4829	85.4829	87.4829	85.9220
17	84.2252	86.2252	88.2252	85.6703
18	77.5356	79.5356	81.5356	79.4226
19	73.7089	75.7089	77.7089	74.9503
20	78.1872	80.1872	82.1872	79.0427
21	87.013	89.013	91.013	89.0594
22	89.4577	91.4577	93.4577	93.4267
23	91.0699	93.0699	95.0699	91.9237
24	91.2642	93.2642	95.2642	93.0951
25	88.8697	90.8697	92.8697	90.8833
26	65.0572	67.0572	69.0572	68.2265
27	66.0277	68.0277	70.0277	67.9805
28	75.0352	77.0352	79.0352	77.0118
29	51.5102	53.5102	55.5102	55.0350
30	51.1251	53.1251	55.1251	51.7301

$$\tilde{Y}_{a=0.4} = 0.6 \begin{bmatrix} (70.4154, 2) + \left(0.2662, 2.89 \times 10^{-15}\right)DO + \left(-1.1072, -8 \times 10^{-15}\right)BOD \\ + \left(0.0709, 5.26 \times 10^{-15}\right)COD + \left(-2.8051, -3.6 \times 10^{-15}\right)AN \\ + (-0.0535, 2.22 \times 10^{-16})SS + (-0.3425, -1.4 \times 10^{-14})pH \end{bmatrix}$$

$$+ 0.4 \begin{bmatrix} (70.4154, 2) + \left(0.2662, 2.89 \times 10^{-15}\right)DO + \left(-1.1072, -8 \times 10^{-15}\right)BOD \\ + \left(0.0709, 5.26 \times 10^{-15}\right)COD + \left(-2.8051, -3.6 \times 10^{-15}\right)AN \\ + (-0.0535, 2.22 \times 10^{-16})SS + (-0.3425, -1.4 \times 10^{-14})pH \end{bmatrix}$$

$$\tilde{Y}_{a=0.6} = 0.4 \begin{bmatrix} (70.4154, 2) + \left(0.2662, 2.89 \times 10^{-15}\right)DO + \left(-1.1072, -8 \times 10^{-15}\right)BOD \\ + \left(0.0709, 5.26 \times 10^{-15}\right)COD + \left(-2.8051, -3.6 \times 10^{-15}\right)AN \\ + (-0.0535, 2.22 \times 10^{-16})SS + (-0.3425, -1.4 \times 10^{-14})pH \end{bmatrix}$$

$$+ 0.6 \begin{bmatrix} (70.4154, 2) + \left(0.2662, 2.89 \times 10^{-15}\right)DO + \left(-1.1072, -8 \times 10^{-15}\right)BOD \\ + \left(0.0709, 5.26 \times 10^{-15}\right)COD + \left(-2.8051, -3.6 \times 10^{-15}\right)AN \\ + (-0.0535, 2.22 \times 10^{-16})SS + (-0.3425, -1.4 \times 10^{-14})pH \end{bmatrix}$$

Figures (1.8 and 1.9) showed the upgrade result from the fuzzy multivariate model and gives good results and is comparable to the previous models. The use of the proposed model can also provide the best classification of WQI ensemble members. Moreover, the use of this model will determine if a single or ensemble approach is suitable to be used. The novel contribution of this model can be used as a guide to predict other rivers WQI. The resulting of classification between the best model of every method is illustrates in Table 1.7.

The error for every models that represents the status of the minimum error are shown in Table 1.8 (Fig. 1.10).

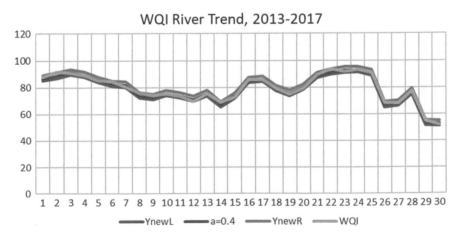

Fig. 1.8 Result for spreading value using fuzzy convex combination for $a = 0.4$

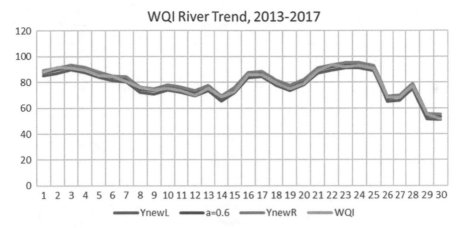

Fig. 1.9   Result for spreading value using fuzzy convex combination for $a = 0.6$

Table 1.7   Class of fuzzy convex combination

River	$a = 0.4$	Y	WQI	Y_{new}	$a = 0.6$
1	II	II	II	II	II
2	II	II	II	II	II
3	II	II	II	II	II
4	II	II	II	II	II
5	II	II	II	II	II
6	II	II	II	II	II
7	II	II	II	II	II
8	III	III	III	III	III
9	III	III	III	III	III
10	III	III	III	III	III
11	III	III	III	III	III
12	III	III	III	III	III
13	III	III	III	III	III
14	III	III	III	III	III
15	III	III	III	III	III
16	II	II	II	II	II
17	II	II	II	II	II
18	II	II	II	II	II
19	III	III	III	III	III
20	II	II	II	II	II
21	II	II	II	II	II
22	II	II	I	II	II
23	II	I	II	I	I

(continued)

Table 1.7 (continued)

River	$a = 0.4$	Y	WQI	Y_{new}	$a = 0.6$
24	I	I	I	I	I
25	II	II	II	II	II
26	III	III	III	III	III
27	III	III	III	III	III
28	II	II	II	II	II
29	III	III	III	III	III
30	III	III	IV	III	III

After several simulation and error measurement, we can conclude that the best model to calculate WQI value is the convex combination method when the free parameter, $a = 0.4$ compared to the other method. This value will give smaller error. Furthermore, the class of WQI obtained from the proposed model is almost the same as WQI value obtain from DOE sampling. The development of fuzzy convex combination model, which includes the fuzzy multivariate regression model has better performance compared with fuzzy multivariate regression.

Table 1.8 Error in fuzzy convex combination

River	$a = 0.4$	$a = 0.6$	Y	Y_{new}
1	1.3127605	0.4027785	0.8577	0.8577
2	3.115833	2.24276907	2.6793	2.6791
3	0.1938137	1.07279966	0.6333	0.6337
4	0.1798527	1.07748689	0.6286	0.6286
5	0.511688	1.45004296	0.9808	0.9808
6	2.1472359	1.20002908	1.6736	1.6736
7	1.6308698	2.62370254	2.1272	2.1272
8	2.9117142	1.85825773	2.3849	2.3849
9	1.621248	0.53751947	1.0793	1.0793
10	0.1722516	1.23369882	0.7029	0.7029
11	0.4364985	1.52545416	0.9809	0.9809
12	1.997857	3.14434919	2.5711	2.5711
13	0.0333569	1.02812071	0.4973	0.4973
14	2.435886	1.27181416	1.8538	1.8538
15	1.1133288	2.21484229	1.6640	1.6640
16	0.9765608	0.0454845	0.5110	0.5110
17	0.1808086	1.11462098	0.6477	0.6477

(continued)

Table 1.8 (continued)

River	$a = 0.4$	$a = 0.6$	Y	Y_{new}
18	0.3614298	0.64583912	0.1422	0.1422
19	0.4784038	1.54577712	1.0120	1.0120
20	0.9419548	1.95406557	1.4480	1.4480
21	0.5012631	0.39701354	0.0521	0.0521
22	2.5356639	1.67937784	2.1075	2.1075
23	0.8117505	1.68203704	1.2468	1.2468
24	0.2479985	0.61133757	0.1816	0.1816
25	0.4550818	0.4251678	0.0149	0.0149
26	2.3000444	1.12747961	1.7137	1.7137
27	0.5190594	0.65774743	0.0693	0.0693
28	0.4890566	0.54974422	0.0303	0.0303
29	3.4973889	2.04376926	2.7705	2.7705
30	1.9234428	3.46992964	2.6966	2.6966

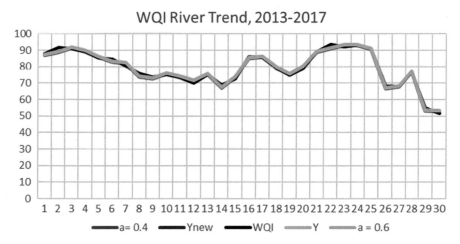

Fig. 1.10 Result for the best spreading value

1.6 Prediction of Water Quality Index

The proposed predictive modelling for WQI comprise of three predictive ranges using the best method from previous experiments and research. The predictive model for WQI used the fuzzy convex combination at $a = 0.4$ for 30 different rivers located in Perak river basin. Table 1.9 summarizes all the results.

The prediction of fuzzy convex combination form for membership function, $\mu = 0$ when $a = 1.5$ and $a = -1.5$, (Fig. 1.11; Table 1.10)

Table 1.9 Fuzzy regression model form for membership function for $a = 0.4$

River	Y_{newL}	WQI		$a = 0.4$		Y_{newR}
1	91.40469	92.51827	II	93.00469	I	95.40469
2	87.75704	91.67735	II	89.35704	II	91.75704
3	91.40875	90.833	II	93.00875	I	95.40875
4	91.35952	92.2532	II	92.95952	I	95.35952
5	81.34083	84.58931	II	82.94083	II	85.34083
6	89.68939	92.32753	II	91.28939	II	93.68939
7	87.15993	92.79588	I	88.75993	II	91.15993
8	88.04936	89.91799	II	89.64936	II	92.04936
9	86.73836	90.31982	II	88.33836	II	90.73836
10	84.5285	87.16483	II	86.1285	II	88.5285
11	83.97912	88.15223	II	85.57912	II	87.97912
12	81.33324	85.50357	II	82.93324	II	85.33324
13	81.5404	86.15981	II	83.1404	II	85.5404
14	78.97825	80.57879	II	80.57825	II	82.97825
15	80.70783	83.92792	II	82.30783	II	84.70783

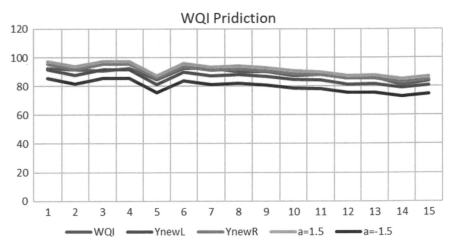

Fig. 1.11 Result for spreading value using fuzzy convex combination for $a = 1.5$ and $a = -1.5$

Table 1.10 WQI list of class prediction for fuzzy convex combination

River	$a = 1.5$		WQI		$a = -1.5$	
1	97.40469	I	92.51827	II	85.40469	II
2	93.75704	I	91.67735	II	81.75704	II
3	97.40875	I	90.833	II	85.40875	II
4	97.35952	I	92.2532	II	85.35952	II
5	87.34083	II	84.58931	II	75.34083	III
6	95.68939	I	92.32753	II	83.68939	II
7	93.15993	I	92.79588	I	81.15993	II
8	94.04936	I	89.91799	II	82.04936	II
9	92.73836	I	90.31982	II	80.73836	II
10	90.5285	II	87.16483	II	78.5285	II
11	89.97912	II	88.15223	II	77.97912	II
12	87.33324	II	85.50357	II	75.33324	III
13	87.5404	II	86.15981	II	75.5404	III
14	84.97825	II	80.57879	II	72.97825	III
15	86.70783	II	83.92792	II	74.70783	III

$$\tilde{Y}_{a=1.5} = -0.5 \begin{bmatrix} (70.4154, 2) + \left(0.2662, 2.89 \times 10^{-15}\right)DO + \left(-1.1072, -8 \times 10^{-15}\right)BOD \\ + \left(0.0709, 5.26 \times 10^{-15}\right)COD + \left(-2.8051, -3.6 \times 10^{-15}\right)AN \\ + (-0.0535, 2.22 \times 10^{-16})SS + (-0.3425, -1.4 \times 10^{-14})pH \end{bmatrix}$$

$$+ 1.5 \begin{bmatrix} (70.4154, 2) + \left(0.2662, 2.89 \times 10^{-15}\right)DO + \left(-1.1072, -8 \times 10^{-15}\right)BOD \\ + \left(0.0709, 5.26 \times 10^{-15}\right)COD + \left(-2.8051, -3.6 \times 10^{-15}\right)AN \\ + (-0.0535, 2.22 \times 10^{-16})SS + (-0.3425, -1.4 \times 10^{-14})pH \end{bmatrix}$$

$$\tilde{Y}_{a=-1.5} = +2.5 \begin{bmatrix} (70.4154, 2) + \left(0.2662, 2.89 \times 10^{-15}\right)DO + \left(-1.1072, -8 \times 10^{-15}\right)BOD \\ + \left(0.0709, 5.26 \times 10^{-15}\right)COD + \left(-2.8051, -3.6 \times 10^{-15}\right)AN \\ + (-0.0535, 2.22 \times 10^{-16})SS + (-0.3425, -1.4 \times 10^{-14})pH \end{bmatrix}$$

$$- 1.5 \begin{bmatrix} (70.4154, 2) + \left(0.2662, 2.89 \times 10^{-15}\right)DO + \left(-1.1072, -8 \times 10^{-15}\right)BOD \\ + \left(0.0709, 5.26 \times 10^{-15}\right)COD + \left(-2.8051, -3.6 \times 10^{-15}\right)AN \\ + (-0.0535, 2.22 \times 10^{-16})SS + (-0.3425, -1.4 \times 10^{-14})pH \end{bmatrix}$$

1.7 Summary

The focus of this study is on the development of fuzzy multivariate regression methods to estimate the value of WQI at selected rivers in the State of Perak, Malaysia. The main contribution of the present study is the enhancement of data preprocessing techniques and the new formulation to calculate spreading value. For data preprocessing, fuzzy convex combination which embedded with fuzzy multivariate regression for given spreading have been proposed. The proposed methods have proven to be able to increase the classification accuracy compared to the existing methods. Furthermore, the predicting models produced in this study includes; fuzzy multivariate regression and convex combination method (weighted) using new spreading formula. The proposed models have been successfully tested and validated with benchmark and real data. Thus, the proposed fuzzy multivariate regression model is the best compared with some existing method.

Acknowledgements This study is fully supported by Universitas Islam Riau (UIR), Pekan baru, Indonesia and Universiti Teknologi PETRONAS (UTP), Malaysia through **International Collaborative Research Funding (ICRF): 015ME0-037**. The first author is currently doing his internship at UTP under Research Attachment Program (RAP).

References

1. L. Abdullah, N. Zakaria, Matrix driven multivariate fuzzy linear regression model in car sales. J. Appl. Sci. **12**, 56–63 (2012)
2. ASMA.: River Water Quality Monitoring (2012). http://www.doe.gov.my/portalv1/en/general-info/pemantauan-kualiti-air-sungai/280 retrieved from 13 May 2019
3. A.A.A. Bakar, A.M. Pauzi, A.A. Mohamed, S.S. Sharifuddin, F.M. Idris, Preliminary analysis on the water quality index (WQI) of irradiated basic filter elements, in *IOP Conference Series: Material Science and Engineering*, vol. 298, 012005 (2018)
4. L.W. Canter, *Environmental Impact of Water Resources Projects* (Lewis Publishers, Inc. Chelsea, 1985)
5. P.T. Chang, E.S. Lee, A generalized fuzzy weighted least-squares regression. Fuzzy Sets Syst. **82**, 289–298 (1996)
6. P. D'Urso, T. Gastaldi, A least-squares approach to fuzzy linear regression analysis. Comput. Stat. Data Anal. **34**, 427–440 (2000)
7. P. D'Urso, T. Gastaldi, Linear fuzzy regression analysis with asymmetric spreads, in *Advances in Data Science and Classification*, ed. by S. Borra, R. Rocci, M. Vichi, M. Schader (Springer, Heidelberg, 2001), pp. 257–264
8. B. Heshmati, A. Kandel, Fuzzy linear regression and its application to forecasting in uncertain environment. Fuzzy Set Syst. **15**, 159–191 (1985)
9. R.K. Horton, An index number system for rating water quality. J. Water Pollut. Control Fed. **37**(3), 300–305 (1965)
10. N.H.M. Isa, M. Othman, S.A.A. Karim, Multivariate matrix for fuzzy linear regression model to analyze the taxation in Malaysia. Int. J. Eng. Technol. **7**(4.33), 78–82 (2018)
11. L.Y. Khuan, N. Hamzah, R. Jailani, Prediction of water quality index (WQI) based on artificial neural network (ANN), in *2002 Student Conference on Research and Development Proceedings* (Shah Alam, Malaysia, 2002), pp. 157–161

12. A.A. Mamun, S.N. Hafizah, M.Z. Alam, Improvement of existing water quality index in Selangor, Malaysia, in *2nd International Conference on Water & Flood Management (ICWFM)*, 15–17 Mar. 09, Dhaka, Bangladesh (2009)
13. W.R. Ott, *Water Quality Indices: A Survey of Indices Used in the United States*, EPA-600/4-78-005 (US Environmental Protection Agency, Washington, 1978), pp. 128
14. N.F. Pan, T.C. Lin, N.H. Pan, Estimating bridge performance based on a matrix-driven fuzzy linear regression model. Autom. Constr. **18**, 578–586 (2009). https://doi.org/10.1016/j.autcon.2008.12.005
15. H.I. Sii, J.H. Sherrard, T.E. Wilson, A water quality index based on fuzzy sets theory, in *Proceedings of the 1993 Joint ASCE-CSCE National Conference on Environmental Engineering, July 12–14, Montreal, Quebec, Canada* (1993), pp. 253–259
16. H. Tanaka, S. Uejima, K. Asai, Linear regression analysis with fuzzy model. IEEE Trans. Syst. Man Cybern. **12**(6), 903–907 (1982)
17. L.A. Zadeh, Fuzzy set. Inform. Control **8**, 338–353 (1965)
18. N.S. Zainordin, N.A. Ramli, M. Elbayoumi, Distribution and temporal behavior of O_3 and NO_2 near selected schools in Seberang Perai, Pulau Pinang and Parit Buntar, Perak. Malaysia. Sains Malaysiana **46**(2), 197–207 (2017)

Chapter 2
Smart Sensor Node of Wireless Sensor Networks (WSNs) for Remote River Water Pollution Monitoring System

Evizal Abdul Kadir, Abdul Syukur, Mahmod Othman, and Bahruddin Saad

Indonesia is one of the countries that have many rivers and lakes. It is situated, in South East Asia and enjoys tropical climate all year round. Riau province is located in the centre and middle of Sumatera Island which in the heart of Sumatera. This province has more than five big rivers that are used by the community every day for their daily activities. The rapid economic development has significant impact to the region where many industries operating along the river produce industrial wastes that pollutes the river water. This chapter discusses the development of river water monitoring system where several relevant parameters are monitored. The Wireless Sensor Networks (WSNs) used in this study integrates a sensor node that is attached to multiple sensors such as water temperature, pH, dissolved oxygen (DO) and electrical conductivity. The monitoring system is specially designed to be able to monitor water level and flow rate for environmental and flood alert purposes. A sensor node collects information from the multiple sensors and forwards it to the WSNs sink node which is embedded with a microcontroller unit and a memory as local database before sending signals to the backend system. The backend system displays vital information that can be monitored by institutions or local authorities. Prompt action will be taken if abnormality is raised by the monitoring system. A prototype of this WSNs node has been designed and tested, the results shows that sensor node is reliable for the detection of polluted water parameters, water levels as well as river flow rate. Furthermore, sensor node was tested at one of the rivers in Riau Province, Indonesia to compare results with actual river water. All the data were kept

E. A. Kadir (✉) · A. Syukur
Department of Informatics, Faculty of Engineering, Universitas Islam Riau, Jl. Kaharuddin Nasution, Marpoyan, Pekanbaru, Riau 28284, Indonesia
e-mail: evizal@eng.uir.ac.id

M. Othman · B. Saad
Fundamental and Applied Sciences Department, Universiti Teknologi PETRONAS, Seri Iskandar, Malaysia

© The Author(s), under exclusive license to Springer Nature Singapore Pte Ltd. 2020   29
S. A. A. Karim et al. (eds.), *Optimization Based Model Using Fuzzy and Other Statistical Techniques Towards Environmental Sustainability*,
https://doi.org/10.1007/978-981-15-2655-8_2

in the database for recording and to do analysis as well as for future development of monitoring system.

2.1 Introduction

In some countries, especially the developing countries, river remains an important part for daily activities such as transportation, as floating home, washing, shower and even for cooking without filtration system. Economic enhancements are boosted by many companies that operate along the river for easy in transportation as well as other considerations. Riau Province in Indonesia has six rivers, one of them being the deepest in Indonesia. The many industries operating along the river caused severe water pollution because of the wastes generated and often the unclean environmental operations. Polluted water may contain abnormal parameters such as pH, dissolved oxygen (DO), temperature, electrical conductivity, etc.

The traditional method to monitor water quality has been done using laboratory-based testing of the collected water samples. Though this method, complete range of laboratory tests including physical, biological, and chemical parameters are possible but it is not practical to measure several points along the river [1–3]. Additionally, laboratory-based tests could take up to several days to achieve the results and for some parameters may show less accuracy because of the sample water changes during sampling. Recently, real-time sensors for environmental monitoring are beginning to gain popularity due to the rapid advancement in sensor technology, especially in WSNs technology that could be adopted in many kinds of application. The continuous collection of water quality data as well as real-time monitoring system can be used to monitor the state of a river ecosystem, establish trends and determine specifics related to event detection [4–6].

Water pollutant monitoring done in previous research is limited to only a few parameters and most of them monitor only basic water parameters [7–9]. Water pollution monitoring system proposed in [10–12] used multi sensors but limited sensor that only cover basic parameter of water which is temperature and pH, as well as the data keep in local makes incompatible to online remote monitoring. Analysis of water quality by image processing and remote sensing for long distance monitoring has accuracy problems [13, 14]. The use of robotics in water quality monitoring of deep rivers and oceans has obvious advantages but the cost is prohibitive and required skillful operators [15–17].

This research aims to develop a new sensor node for WSNs system with the ability to obtain multiple water quality parameters at one of the rivers in Riau Province, Indonesia. Beside the real-time based monitoring, the system includes river water level and river flow rate sensors, parameters that are vital for flood managements during rainy season. The research contributes to new knowledge and offer invention for water quality monitoring system by data collection, including a new sensor design that is able to collect accurate data. A new method of communication system from sensor nodes to WSNs gateway through WSN sink for effective data transmission and sharing is also an important aim of this research. With the local or remote data monitoring center, complete monitoring of interface implemented to obtain historical data queries, real-time data and network state display, data analysis and alarm for abnormal situations is made possible.

2.2   Sensor Node Design for Water Monitoring

The design of sensor node for WSNs system application for river water monitoring system is based on early data that was collected by analysing river water samples. Based on results of sample river water, then a smart node sensor for WSN system to approach pollution parameters and material contamination to the river water was designed. Figure 2.1 shows a scene from a part of river in Riau Province Indonesia, where the community uses river and river water as daily activities such as for transportation, shower, washing and even for cooking.

The location to conduct monitoring at Siak river is Riau province, Indonesia and the actual location as shows in Fig. 2.2 for the latitude and longitude coordinates are 0.539062, 101.445321.

The actual condition of the river and river water in Riau Province, Indonesia is very bad and poses high risk to the ecosystem along the river as well as for humans and community that use river water for their daily activities. Figure 2.3 shows an actual condition of polluted river that was contaminated by material or chemicals

Fig. 2.1 A scene from a river in Riau Province, Indonesia

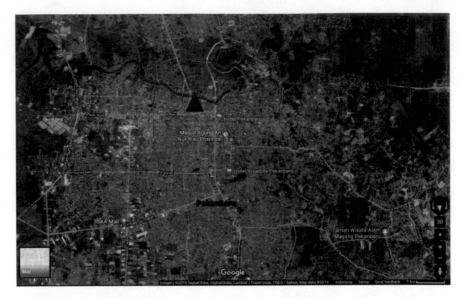

Fig. 2.2 A map where the location to do the monitoring

Fig. 2.3 Water polluted by chemicals from an industry operating along river

Table 2.1 Sensor node design specification

Parameter	Range	Accuracy	Method
Temperature	0–16 °C	±0.5 °C	Thermistor
DO	0–20 mg/L	±0.5 mg/L	Polarography
pH	0–14	±0.1	Glass Electrode
Salinity	0–50%	±0.5	Conductivity
			Measurement
Flow rate	0–10 m³/s	±0.1	Flow sensor
Water level	−5 to 10 m	±0.1	Level meter

from an industry operating beside the river (circle bottom left), a few kids swimming and playing in the river can be seen at top right of the Fig. 2.3. Based on these observations and water analysis, some parameters or indicators of water quality is urgently required to be monitored such as temperature, pH, dissolved oxygen (DO) and electrical conductivity. The river water monitoring system was designed as not only for water pollution monitoring, furthermore to make system can be added sensor node for future development. Additionally, water level and flow rate measurements are important indicators of flooding. Most of river located in Riau are at a very high risk of flooding, especially during the rainy season. Flooding alert system is very important to remind the community when water level has reach dangerous level:

Smart sensor node of WSNs consists of six parameters as indicators to measure water pollutant and river water status alert. Table 2.1 shows the complete measurement indicator with range of sensor as well as the accuracy.

Figure 2.4 shows a block diagram of the smart sensor node for WSNs system every data collected by the sensor unit will be stored in a local database, then all the analyzed data will be forwarded to data center at the backend system.

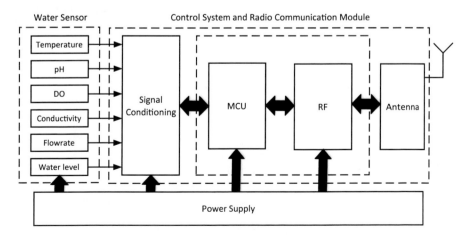

Fig. 2.4 Block diagram of smart sensor node for WSNs system

2.3 River Water Pollutant Monitoring System

The system possesses typical WSNs structure with a novel design of sensor nodes, which is easily configured as an arbitrary parameter or multi-parameter monitoring networks. Based on comparison with the manual river water monitoring system, [18] it has the following merits: (i) The sensor nodes are attached with multiple sensors and low power with independent power supply through solar panel system. (ii) The monitoring parameters are flexible; the sensor network on the monitored area is self-organized, (iii) the capacity of network is big, and the node distribution can be much denser. All the information will be shared to the community. A display of information about water quality will be installed at the community centre and all the people can have access to information including river water levels.

2.3.1 WSNs Sensing System

A set of sensing system with all the sensors to detect how much river was contaminated was installed at the river side in order to obtain, actual data on the river flow. Figure 2.5 illustrates a sensor node that was installed on the river with independent power supply system from solar. The sensor node is normally installed very far from the city area; thus, normal power supply is not available. Thus, the solar powered system with backup battery become very handy.

Large quantity of raw data is collected from every sensor contribute large quantity, since the sensor node has limited data storage, the large data resulted in slow response while sending data to the sink node (gateway). Multiple sensor will affect the sensor node performance as well, thus an intelligent sensor node designed to achieve quick response when abnormal detection on river water monitoring was introduced. Introducing algorithm to the sensor node and some data filtering to avoid waste of data, enables the sensor node to become intelligent and smart in the detection of river water pollution.

2.3.2 WSNs Sink Node and Communication System

In order to achieve accurate monitoring data from the sensor node installed and because the river quite long (more than 50 km), thus a few sensor nodes have to be installed. The average distance between the nodes is different based on early data collected by geographical survey of the river as well as the number of the industries operating along the river. Furthermore, community village and some other activities have contributed to river water pollution, to obtain more accurate data, the average node distance must be installed as close as possible (less than 1 km). Figure 2.6 show a scenario of network topology for the sensing system with a number of sensor nodes,

Fig. 2.5 WSNs node installed at the river side to detect water pollution

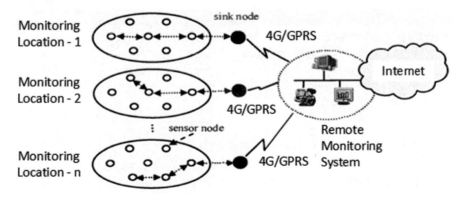

Fig. 2.6 Sensor node communication to sink and backend system

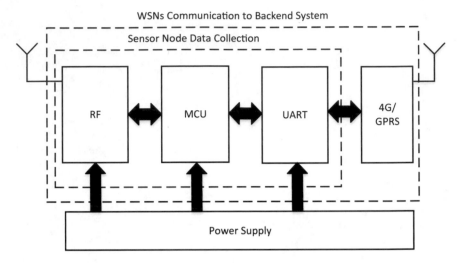

Fig. 2.7 Block diagram of sink node to communicate to sensor node

where every sensor node has their own sink node for collecting data in local host before forwarding to the monitoring center (backend system). In this case, Fourth Generation (4G) network was proposed for sink node communication to the data center for faster communication and real-time monitoring, as currently most of area is covered by 4G network.

Real-time data is required to obtain fast response in the event that the river water get polluted by chemicals or other materials.

Figure 2.7 shows a block diagram of a WSNs sink node as gateway to communicate between sensor node for data collection and to the monitoring center. In this design, the sensing node was able to serve up to 50 sensor nodes where all the information was stored in the buffer before forwarding to the backend system. Many types of wireless communication system to transfer data from sink node monitoring system is available but, in this case, 4G wireless network was used for faster data transfer since the sink node is located very far away from the data center. Dedicated wireless communication system is applicable to use but required line of sight (LOS) and involves high cost to setup.

2.4 Simulation Results and Discussion

Results of simulation is done in the laboratory testing; these results were used as initial data before the actual sensor was installed. The initial results were very useful to check whether the proposed sensor model and type is good and applicable to be used in sensor node based on design of parameter set. Some of the data were compared to other sensor data sheet and literature as reference [5]. Results of the

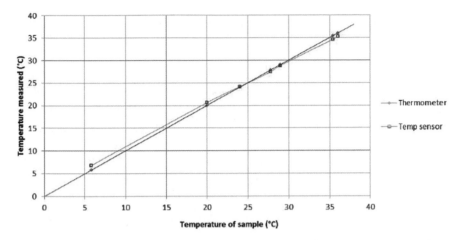

Fig. 2.8 Temperature sensor results versus thermometer

temperature sensor tested were compared to thermometer (Fig. 2.8). The thermometer temperature measurements were compared to temperature sensor assembly. The average difference between the temperature sensors reading to thermometer reading is between min 0.071 °C and maximum 1 °C.

Conductivity is one of the parameters measured in water pollutant index, based on a 2-electrode method design. The sensor was expected to provide accuracy of at least 15%. The initial test was conducted to observe the accuracy of the signal conditioning circuit. Figure 2.8 shows the signal conditioning test results compared to the theoretically simulated conductivity.

Water pH is another very important parameter to check quality of the river water. The sensor used for sensor node was based on the glass electrode. The pH sensor was designed with specification and accuracy of at least 0.4 pH. Two classifications of tests were conducted to observe the accuracy of the installed pH sensor. Figure 2.9 shows pH sensor when tested between measurement and theoretical analysis.

Water flow sensor is used to measure flow rate of the river, because flow rate of liquid is related to pressure and depth of the river. The flow rate sensor was selected based on the river depth and width to design a flow meter. The signal received from the sensor was analyzed bit by bit and converted to the unit then the value of river water flow rate is obtained. The signal conditioning received from sensor was mostly based on the amplitude of the voltage pulses. Figure 2.10 shows an initial test of water flow rate and measurement results.

Based on initial test in the laboratory and measurement of all the sensors, good results and good agreement between theoretical and actual measurements were obtained. Further installation and test to the actual site is required to check and measure in actual site conditions with under environments. Figure 2.11 shows the graph of water flow rate based on test conducted.

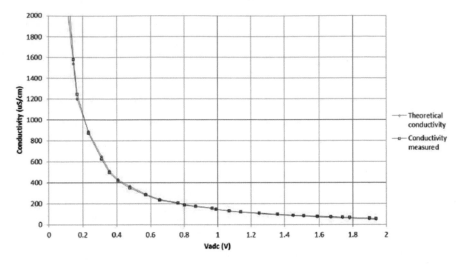

Fig. 2.9 Electrical conductivity test results for sensor node

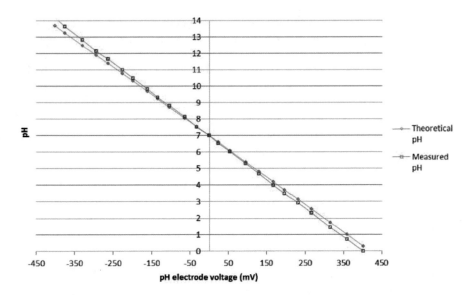

Fig. 2.10 Water pH sensor test between theoretical and actual measurements

2.5 Conclusion

The design of smart sensor for WSNs node have been proposed with multiple sensors to measure all the parameters in the polluted river water. Based on initial laboratory sample test of river water, since there are many parameters and chemicals that were involved, thus various sensors such as water temperature meter were used. Water pH

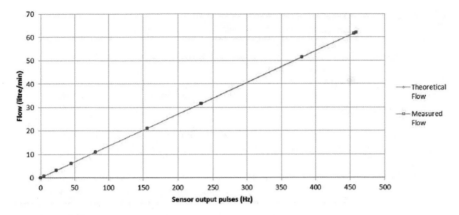

Fig. 2.11 Water flow rate sensor testing compare to theoretical

is one of the parameters monitored as well as water DO. Electrical conductivity is measured in order to detect waste from industry or other pollutants. Furthermore, to make sensor node beneficial for the community, a set of sensors, i.e., water level and water flow rate sensors were installed in the sensor node to measure river water status as alert for in the event of floods. Measurement results show good agreement compared to the theoretical analysis for all the sensors. Thus, the sensor node can be applied and ready to be deployed to actual sites. This approach was applied to avoid dump and waste data in the sink node as well as for wireless data communication to the backend system.

Acknowledgements Authors would like to thank to Universiti Teknologi PETRONAS, Malaysia and KEMENRISTEKDIKTI Indonesia for funding this research project and Universitas Islam Riau, Indonesia for supporting all the facilities.

References

1. S. Zhuiykov, Solid-state sensors monitoring parameters of water quality for the next generation of wireless sensor networks. Sens. Actuators B Chem. **161**(1), 1–20 (2012)
2. T.P. Lambrou, C.G. Panayiotou, C.C. Anastasiou, A low-cost system for real time monitoring and assessment of potable water quality at consumer sites, in *IEEE Conference on Sensor*, Taipei, Taiwan, (2012), pp. 1–4
3. A. Aisopou, I. Stoianov, N. Graham, In-pipe water quality monitoring in water supply systems under steady and unsteady state flow conditions: a quantitative assessment. Water Res. **46**(1), 235–246 (2012)
4. L.Y. Li, H. Jaafar, N.H. Ramli, Preliminary study of water quality monitoring based on WSN technology, in *2018 International Conference on Computational Approach in Smart Systems Design and Applications (ICASSDA)* (2018), pp. 1–7
5. N.A. Cloete, R. Malekian, L. Nair, Design of smart sensors for real-time water quality monitoring. IEEE Access **4**, 3975–3990 (2016)

6. E.A. Kadir, S.L. Ros, A. Yulianti, Application of WSNs for detection land and forest fire in Riau Province Indonesia, in *International Conference on Electrical Engineering and Computer Science (ICECOS)*, Bangka Belitung (IEEE, 2018)

7. T.P. Lambrou, C.C. Anastasiou, C.G. Panayiotou, M.M. Polycarpou, A low-cost sensor network for real-time monitoring and contamination detection in drinking water distribution systems. IEEE Sens. J. **14**(8), 2765–2772 (2014)

8. J. Tian, Y. Wang, A novel water pollution monitoring approach based on 3 s technique, in *2010 International Conference on E-Health Networking Digital Ecosystems and Technologies (EDT)*, vol. 1 (2010), pp. 288–290

9. M. Grossi, R. Lazzarini, M. Lanzoni, A. Pompei, D. Matteuzzi, B. Riccò, A portable sensor with disposable electrodes for water bacterial quality assessment. IEEE Sens. J. **13**(5), 1775–1782 (2013)

10. S. Randhawa, S.S. Sandha, B. Srivastava, A multi-sensor process for in-situ monitoring of water pollution in rivers or lakes for high-resolution quantitative and qualitative water quality data, in *2016 IEEE Intl Conference on Computational Science and Engineering (CSE) and IEEE Intl Conference on Embedded and Ubiquitous Computing (EUC) and 15th Intl Symposium on Distributed Computing and Applications for Business Engineering (DCABES)* (2016), pp. 122–129

11. T. Li, M. Xia, J. Chen, Y. Zhao, C. de Silva, Automated water quality survey and evaluation using an IoT platform with mobile sensor nodes. Sensors **17**(8), 1735 (2017)

12. M. Cheng, Z. Guo, H. Dang, Y. He, G. Zhi, J. Chen, Y. Zhang, W. Zhang, F. Meng., Assessment of the evolution of nitrate deposition using remote sensing data over the Yangtze River Delta, China. IEEE J. Sel. Top. Appl. Earth Obs. Remote Sens. **9**(8), 3535–3545 (2016)

13. C. Doña, J.M. Sánchez, V. Caselles, J.A. Domínguez, A. Camacho, Empirical relationships for monitoring water quality of lakes and reservoirs through multispectral images. IEEE J. Sel. Top. Appl. Earth Obs. Remote Sens. **7**(5), 1632–1641 (2014)

14. S. Olatinwo, T.-H. Joubert, Optimizing the energy and throughput of a water-quality monitoring system. Sensors **18**(4), 1198 (2018)

15. P. Teixidó, J. Gómez-Galán, F. Gómez-Bravo, T. Sánchez-Rodríguez, J. Alcina, J. Aponte, Low-power low-cost wireless flood sensor for smart home systems. Sensors **18**(11), 3817 (2018)

16. F.D.V.B. Luna, E.d.l.R. Aguilar, J.S. Naranjo, J.G. Jagüey, Robotic system for automation of water quality monitoring and feeding in aquaculture Shadehouse. IEEE Trans. Syst. Man Cybern. Syst. **47**(7), 1575–1589 (2017)

17. Z. Wu, J. Liu, J. Yu, H. Fang, Development of a novel robotic Dolphin and its application to water quality monitoring. IEEE/ASME Trans. Mechatron. **22**(5), 2130–2140 (2017)

18. N. Maojing, River water quality monitoring and simulation based on WebGIS—Anhui Yinghe River as an Example, in *2016 Sixth International Conference on Instrumentation & Measurement, Computer, Communication and Control (IMCCC)* (2016), pp. 716–720

Chapter 3
Toward Formalization of Comprehensive Bilingual Dictionaries Creation Planning as Constraint Optimization Problem

Arbi Haza Nasution, Evizal Abdul Kadir, Yohei Murakami, and Toru Ishida

Bilingual lexicon extraction is problematic for low-resource languages due to the paucity of parallel and comparable corpora. The pivot-based induction techniques have been proven useful for inducing bilingual lexicons with only bilingual dictionary as input. When we want to create multiple bilingual dictionaries linking several languages, we need to consider manual creation by human if there are no machine-readable dictionaries available as input. We formalize a comprehensive bilingual dictionaries creation planning that utilizes four pivot-based induction techniques and a manual creation as constraint optimization problem with an objective to reduce total cost.

3.1 Closely Related Languages

Historical linguistics is the scientific study of language change over time in term of sound, analogical, lexical, morphological, syntactic, and semantic information [1].

A. H. Nasution (✉) · E. A. Kadir
Department of Informatics Engineering, Faculty of Engineering, Universitas Islam Riau, Pekanbaru, Indonesia
e-mail: arbi@eng.uir.ac.id

Y. Murakami
Faculty of Information Science and Engineering, Ritsumeikan University, Kyoto, Japan

T. Ishida
School of Creative Science and Engineering, Waseda University, Tokyo, Japan

© The Author(s), under exclusive license to Springer Nature Singapore Pte Ltd. 2020
S. A. A. Karim et al. (eds.), *Optimization Based Model Using Fuzzy and Other Statistical Techniques Towards Environmental Sustainability*,
https://doi.org/10.1007/978-981-15-2655-8_3

Comparative linguistics is a branch of historical linguistics that is concerned with language comparison to determine historical relatedness and to construct language families [2]. Many methods, techniques, and procedures have been utilized in investigating the potential distant genetic relationship of languages, including lexical comparison, sound correspondences, grammatical evidence, borrowing, semantic constraints, chance similarities, and sound-meaning isomorphism [3]. The genetic relationship of languages is used to classify languages into language families. Closely-related languages are those that came from the same origin or proto-language, and belong to the same language family.

Swadesh List is a classic compilation of basic concepts for the purposes of historical-comparative linguistics. It is used in lexicostatistics (quantitative comparison of lexical cognates) and glottochronology (chronological relationship between languages). There are various version of Swadesh list with a number of words equal 225 [4], 215 and 200 [5], and lastly 100 [6]. To find the best size of the list, Swadesh [7] states that "The only solution appears to be a drastic weeding out of the list, in the realization that quality is at least as important as quantity. Even the new list has defects, but they are relatively mild and few in number."

A widely-used notion of string/lexical similarity is the edit distance or also known as Levenshtein Distance (LD): the minimum number of insertions, deletions, and substitutions required to transform one string into the other [8]. For example, LD between "kitten" and "sitting" is 3 since there are three transformations needed: kitten $\rightarrow$ sitten (substitution of "s" for "k"), sitten $\rightarrow$ sittin (substitution of "i" for "e"), and finally sittin $\rightarrow$ sitting (insertion of "g" at the end).

There are a lot of previous works using Levenshtein Distances such as dialect groupings of Irish Gaelic [9] where they gather the data from questionnaire given to native speakers of Irish Gaelic in 86 sites. They obtain 312 different Gaelic words or phrases. Another work is about dialect pronunciation differences of 360 Dutch dialects [10] which obtain 125 words from Reeks Nederlandse Dialectatlassen. They normalize LD by dividing it by the length of the longer alignment. Tang et al. [11] measure linguistic similarity and intelligibility of 15 Chinese dialects and obtain 764 common syllabic units. Petroni [12] define lexical distance between two words as the LD normalized by the number of characters of the longer of the two. Wichmann et al. [13] extend Petroni definition as LDND and use it in Automated Similarity Judgment Program (ASJP).

The ASJP, an open source software was proposed by Holman [14] with the main goal of developing a database of Swadesh lists [7] for all of the world's languages from which lexical similarity or lexical distance matrix between languages can be obtained by comparing the word lists. The classification is based on 100-item reference list of Swadesh [6] and further reduced to 40 most stable items [15]. The item stability is a degree to which words for an item are retained over time and not replaced by another lexical item from the language itself or a borrowed element. Words resistant to replacement are more stable. Stable items have a greater tendency to yield cognates (words that have a common etymological origin) within groups of closely related languages. The comparison of those researches using levenshtein distances is shown in Table 3.1.

Table 3.1 Comparison of previous works using Levenshtein distances

Topic	Method	Result
Dialect groupings of Irish Gaelic [9]	Gather the data from questionnaire given to native speakers of Irish Gaelic in 86 sites	312 different Gaelic words or phrases
Dialect pronunciation differences of 360 Dutch dialects [10]	Normalize LD by dividing it by the length of the longer alignment	125 words from Reeks Nederlandse Dialectatlassen
Linguistic similarity and intelligibility of 15 Chinese dialects [11]	Measure linguistic similarity with LD	764 common syllabic units
Petroni [12] define LD normalization extension	Define lexical distance between two words as the LD normalized by the number of characters of the longer of the two	LD normalization extension
Wichmann et al. [13] extend Petroni's definition as LDND	The classification is based on 100-item reference list of Swadesh [6] and further reduced to 40 most stable items [15]	Automated similarity judgment program (ASJP)

3.2 Bilingual Dictionary Creation

In this section, we will discuss about the recent bilingual dictionary creation from comparable corpora and some inexpensive pivot-based approaches that only require multilingual bilingual dictionaries as input, and methods utilizing language characteristics.

3.2.1 Extraction from Comparable Corpora

The bilingual lexicon induction approach utilizing comparable corpora depends on the assumption that the term and its translation appear in similar contexts [16, 17] which means that a translation equivalent of a source word can be found by identifying a target word with the most similar context vector in a comparable corpus. Identification of good similarity metrics as signals of translation equivalence is the main research challenge in this area. Recently, Irvine [18] presented a discriminative model of bilingual lexicon induction that significantly outperforms previous models. Their model is capable of combining a wide variety of features/signals that weakly identify translation equivalence when being used individually. When feature weights are discriminatively set, these signals produce dramatically higher translation quality than previous approaches that combined signals in an unsupervised fashion (e.g., using minimum reciprocal rank) where bilingual lexicon induction is considered as a binary classification problem; a pair of source and target language words are predicted as translations of one another or not. For a given source language word, all target language candidates were separately scored and then ranked. A variety of signals [16, 17, 19, 20] derived from source and target monolingual corpora were used as features and a supervision was used to estimate the strength of each.

All of the individual signals of translation equivalence are weak indicators by themselves, while combining diverse signals increases the translation accuracy [18] which can be observed even using a simple baseline combination method like mean reciprocal rank. Their discriminative approach to combining the signals achieves dramatically improved performance. They used seed dictionary to empirically weight the contributions of the different signals. In this section, their 6 signals of translation equivalence are discussed.

3.2.2   Pivot-Based Induction Approach

An intermediate/pivot language approach has been applied in machine translation [21] and service computing [22] researches. In this section, three pivot-based induction approaches are discussed which are the traditional inverse consultation method that only requires bilingual dictionaries, the constraint-based approach that treat bilingual lexicon induction problem as constraint satisfaction problem, and finally a recent work that use WordNets as intermediate resource to generate a new bilingual dictionary.

The first work on bilingual lexicon induction to create bilingual dictionary between language A and language C via pivot language B is Inverse Consultation (IC) [23] by utilizing the structure of input dictionaries to measure the closeness of word meanings and then use the results to prune erroneous translation pair candidates, taking into account that many world languages, especially the low-resource languages are still lack of language resources such as parallel corpora and comparable corpora. The IC approach identifies equivalence candidates (EC) of language A words in language C by looking up language A words in dictionary A-B and then looking up the resulting B words in dictionary B-C. These C words equivalence candidates will be looked up in the inverse dictionary C-B and the resulting B words, i.e., Selection Area (SA) are compared to the B equivalences from dictionary A-B (one-time inverse consultation), and then further the B words SA can be looked up in the inverse dictionary B-A to obtain A words SA to be compared again to the original A words (two-times inverse consultation). For example, in Fig. 3.1, a Japanese word 競争 is looked up in dictionary Japanese-English and the resulting English equivalences (E) are *competition*, *contest*, and *race*. These English equivalences are further looked up in dictionary English-French to obtain the equivalence candidates, which are *compétition*, *concours*, *course*, *race*, and *hate*. *Race* and *hate* are further considered as irrelevant ECs since the pivot English word *race* falls into the following cases:

Fig. 3.1 Inverse consultation method

1. The pivot word has multiple meaning with the same spelling. (in this example, the English word race has two meanings: to compete and human race
2. The pivot word has wider meaning than the source word. (in this example, the English word *race* has wider meaning to *hurry* which Japanese word 競争does not have
3. There are mistakes in the dictionaries.

In Fig. 3.1, the equivalence candidates *compétition* is looked up in inverse dictionary French-English to obtain selection area which consist of *contest, competition*, and *match*. Then the English words in SA are compared with English words in equivalences (E) of Japanese word 競争. The number of elements in common between SA and E indicate the nearness of the meaning between the EC and the original word.

To analyze the method used to filter wrong translation pair candidates induced via the pivot-based approach, Saralegi [24] explored distributional similarity measure (DS) in addition to IC. The analysis showed that IC depends on significant lexical variants in the dictionaries for each meaning in the pivot language, while DS depends on distributions or contexts across two corpora of the different languages. Their analysis also showed that the combination of IC and DS outperformed each used individually.

There are many prior work extending the IC method by utilizing various other language resources such as semantic classes and part-of-speech information [25, 26] WordNet [27], monolingual corpora [28], etc. There are also various techniques proposed to better identify the equivalence candidates. Sjobergh [29] compared full definitions in order to detect words corresponding to the same sense. However, not all the dictionaries provide this kind of information. Mausam [30] researched the use of multiple languages as pivots by using Wiktionary for building a multilingual lexicon, with hypothesis that the more languages used, the more evidences will be found to find translation equivalences. Tsunakawa [31] used parallel corpora to estimate translation probabilities between possible translation pairs and setting a minimum threshold to accept equivalence candidates as correct translations. However, even if this approach achieves the best results, it is not adequate to be implemented to low-resource languages due to the scarcity of the parallel corpora.

Even though the IC method seems very suitable for low-resource languages, especially when dictionaries are the only language resource required, unfortunately, for some low-resource languages, it is often difficult to find machine-readable inverse dictionaries and corpora to filter the wrong translation pair candidates. Moreover, since IC relies on pivot language synonyms to identify correct translations, if the relatively rare used meanings had not existed or was missing from the input bilingual dictionaries, IC would not have been able to detect the correct translations. This may result in low recall.

3.2.3 Constraint-Based Induction Approach

The pivot-based approach is very suitable for low-resource languages, especially when dictionaries are the only language resource required. Unfortunately, for some low-resource languages, it is often difficult to find machine-readable inverse dictionaries and corpora to identify and eliminate the erroneous translation pair candidates. To overcome this limitation, Wushouer et al. [32] proposed to treat pivot-based bilingual lexicon induction as an optimization problem. The assumption was that lexicons of closely-related languages offer instances of one-to-one mapping and share a significant number of cognates (words with similar spelling/form and meaning originating from the same root language).

This assumption yielded the development of a constraint optimization model to induce an Uyghur-Kazakh bilingual dictionary using Chinese language as the pivot, which means that Chinese words were used as intermediates to connect Uyghur words in an Uyghur-Chinese dictionary with Kazakh words in a Kazakh-Chinese dictionary. The proposal uses a graph whose vertices represent words and edges indicate shared meanings; following Soderland [30] it was called a transgraph.

The proposal proceeds as follows: (1) use two bilingual dictionaries as input, (2) represent them as transgraphs where w_1^A and w_2^A are non-pivot words in language A, w_1^B and w_2^B are pivot words in language B, and w_1^C, w_2^C and w_3^C are non-pivot words in language C, (3) add some new edges represented by dashed edges based on the one-to-one assumption, (4) formalize the problem into conjunctive normal form (CNF) and use the Weighted Partial MaxSAT (WPMaxSAT) solver [33] to return the optimized translation results, and (5) output the induced bilingual dictionary as the result. These steps are shown in Fig. 3.2.

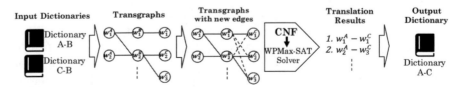

Fig. 3.2 Constraint-based bilingual dictionary induction

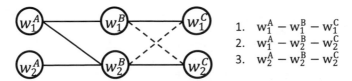

1. $w_1^A - w_1^B - w_1^C$
2. $w_1^A - w_2^B - w_2^C$
3. $w_2^A - w_2^B - w_2^C$

(a) Translation pair candidates from **existing solid-edges** in 1-cycle symmetry assumption

extended to

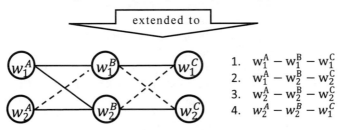

1. $w_1^A - w_1^B - w_1^C$
2. $w_1^A - w_2^B - w_2^C$
3. $w_2^A - w_2^B - w_2^C$
4. $w_2^A - w_2^B - w_1^C$

(b) Translation pair candidates from **existing solid edges** and *previously added dashed-edges* in 2-cycle symmetry assumption

Fig. 3.3 Example of n-cycle symmetry assumption extension

However, the assumption of one-to-one mapping is too strong to induce the many translation pairs needed to offset resource paucity because few such pairs can be found. Therefore, we generalized the constraint-based bilingual lexicon induction framework by extending constraints and translation pair candidates from the one-to-one approach to attain more voluminous bilingual dictionary results with many-to-many translation pairs extracted from connected existing and new edges [34]. We further enhance our generalized method by setting two steps to obtaining translation pair results. First, we identify one-to-one cognates by incorporating more constraints and heuristics to improve the quality of the translation result. We then identify the cognates' synonyms to obtain many-to-many translation pairs. As shown in Fig. 3.3, in each step, we can obtain more cognate and cognate synonym pair candidates by iterating the n-cycle symmetry assumption until all possible translation pair candidates have been reached [35].

3.2.4 Cognates Recognition Approach

Cognates are words with similar spelling/form and meaning that have a common etymological origin. For instance, the words *night* (English), *nuit* (French), *noche* (Spanish), *nacht* (German) and *nacht* (Dutch) have the same meaning which is "night" and derived from the Proto-Indo-European *$n'ok^w ts$* with the same meaning of "night".

The closer the etymological relation between languages, the bigger number of cognates can be found [36]. Since most linguists believe that lexical comparison alone is not a good way to recognize cognates [1], a more general and basic characteristic of closely-related languages need to be utilized such as cognate pair mostly maintain the semantic or meaning of the lexicons. Even though there is a possibility of a change in one of the meanings of a word in a language, within the families where the languages are known to be closely-related, the possibility of a change is smaller.

By utilizing linguistic information, the semantic distance between two words from the word translation topology based on structure of input bilingual dictionaries can be measured. The first step is to recognize one-to-one cognates in the transgraph which share all their senses. Once a list of cognates is obtained, the next step is to recognize cognate synonyms in the transgraph; those that share part/all senses with the cognate and so are mutually connected to some/all pivot words. Those two steps are easy tasks when the input dictionaries have sense/meaning information as shown in Fig. 3.4 where a cognate pair (w_1^A, w_1^C) share two senses, i.e., s_1 and s_2 through pivot word w_1^B and a cognate pair (w_2^A, w_2^C) only share s_1 through pivot word w_1^B and w_2^B. Since for low-resource languages, a machine-readable bilingual dictionary with sense information is scarce, predicting the shared sense/meaning is a research challenge.

Some linguistic studies show that the meaning of a word can be deduced via cognate synonym [37, 38]. For instance, in Fig. 3.4, w_1^A, w_2^A and w_3^A are words in Minangkabau language (min), w_1^B, w_2^B and w_3^B are words in Indonesian language (ind) and w_1^C, w_2^C and w_3^C are words in Malay language (zlm). When we connect words in non-pivot language A and C via pivot words B based on shared meaning between the words, we can get translation results from language A to C. In this example, we have information about senses/meanings for all words in input dictionaries and there are three cognates which are (w_1^A, w_1^B, w_1^C), (w_2^A, w_2^B, w_2^C), and (w_3^A, w_3^B, w_3^C), as indicated within the same box in Fig. 3.5. A cognate w_1^A-w_1^C and non-cognates w_1^A-w_2^C and w_1^A-w_3^C are correct translations since w_1^C, w_2^C and w_3^C are synonymous.

Fig. 3.4 Example of cognate recognition with sense information

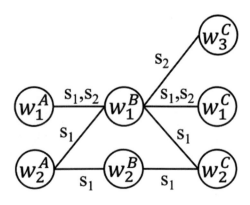

Fig. 3.5 Example of inducing translation pair via pivot words

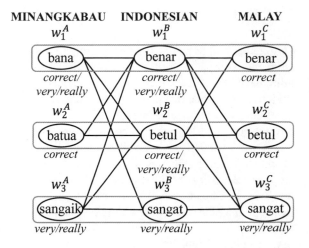

Nevertheless, it remains a challenge to find the cognate synonyms when the input dictionaries do not have information about senses/meanings. Therefore, majority of researches on bilingual dictionary creation of intra-family languages have been put high effort on approaches to detecting cognates. There are two approaches from prior work in detecting cognates. The first approach is to make a description on orthography changes of words. In other word, the researcher want to analyze how orthography of a borrowed word should change when it has been introduced into another language. A work [39] expanded a list of English-German cognate words by applying well-established transformation rules (e.g. substitution of k or z by c and of $-tat$ by $-ty$, as in German *Elektizitat*—English *electricity*). The second approach measure the spelling similarity (also known as form similarity and orthographic similarity) between the given two words. The most well-known approach to measure spelling is edit distance—also known as *Levenshtein distance*—which corresponds to the minimum number of edit operations such as substitution, deletion and insertion required to transform one word into another [8]. A prior work [36] of cognate recognition using edit distance induce bilingual dictionaries between cross-family languages via an intra-family pivot language. The identified cognate pairs are considered as correct translations. Other related techniques of measuring spelling similarity are the longest common subsequent ratio, which counts the number of letters shared by two strings divided by the length of the longest string [40], and another method [41] compare two words by calculating the number of matching consonants. A further extension has been proposed [42], in which authors claimed the importance of genetic cognates by comparing the phonetic similarity of lexemes with the semantic similarity of the glosses.

3.3 Plan Optimization

Despite the higher potential of the pivot-based approach in enriching low-resource languages compared to other methods discussed due to the only input required is bilingual dictionaries, when actually implementing the inverse consultation method [23], the one-to-one mapping approach [32], the constraint-based induction method [34], and the generalized constraint-based induction method utilizing cognate recognition technique [35], we also need to consider the inclusion of a more traditional method like manually creating the bilingual dictionaries by bilingual native speakers. In spite of the high cost, this will be unavoidable if no machine-readable dictionaries are available as input for those methods. Given the various methods and costs that may need to be considered, we recently introduced a plan optimizer to find the feasible optimal plan of creating multiple bilingual dictionaries with the least total cost [43]. The plan optimizer should decides which bilingual dictionary to be invested first or induced right from the start in order to obtain all possible combination of bilingual dictionaries with a satisfying size from the language set with the minimum total cost to be paid.

3.3.1 Motivating Scenario

In order to illustrate the needs of optimal plan for creating multiple bilingual dictionaries with the least total cost we present an example motivating scenario as shown in Fig. 3.6. Consider a stakeholder has a motivation to obtain all 10 combination of bilingual dictionaries from 5 languages with a minimum size of 2000 translation pairs each. Currently, the stakeholder already has a bilingual dictionary of language 1 and 3 $d_{(1,3)}$ with 2100 translation pairs and two bilingual dictionaries $d_{(1,2)}$ and $d_{(2,3)}$ with a number of translation pairs below 2000. Obviously, the stakeholder can just hire native speakers to create and evaluate the bilingual dictionaries following the traditional investment plan to reach his goal with a total cost of C. However, he

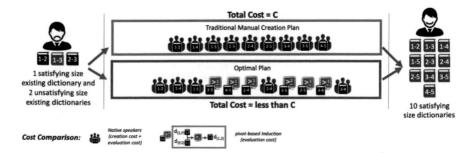

Fig. 3.6 Motivating example

can save cost of bilingual dictionary creation by utilizing several pivot-based bilingual lexicon induction methods with a zero creation cost. Even though the resulting bilingual dictionary still needs to be evaluated by native speakers, by following the optimal plan, the stakeholder can cut significant amount of creation cost.

At this point, the reader might wonder that even before executing the optimal plan, how can we know that utilizing the pivot-based bilingual lexicon induction methods to enrich $d_{(2,3)}$ resulting a satisfying size bilingual dictionary above 2000 translation pairs or below 2000 translation pairs that need to be invested more by native speakers to fill in the gap? To answer this question, the pivot-based bilingual lexicon induction precision need to be estimated in order to calculate the resulting size bilingual dictionary. This uncertainty is the research challenge that need to be addressed in our future work. If one tries to utilize all of the pivot-based bilingual lexicon induction methods and manual creation by native speakers and try to create the plan (order of dictionary creation task to take) manually, the total cost might be higher than following the optimal plan. Since the created bilingual dictionary can be used as input for inducing the other unsatisfying size dictionary, the order of dictionary creation task to take is crucial.

3.3.2 Formalizing Plan Optimization

The plan optimization to bilingual dictionary induction involves discovering the order of bilingual dictionary creation task from a set of possible tasks including the inverse consultation method [23], the one-to-one mapping approach [32], the constraint-based induction method [34], the generalized constraint-based induction method utilizing cognate recognition technique [35], and the manual creation by native speakers to minimize the total cost. We assume that the number of existing translation pairs for existing bilingual dictionaries and the minimum number of translation pairs the output bilingual dictionary should have are both known. Multiple candidate plans exist to finally obtain all bilingual dictionaries. One criteria for selecting a plan is to establish a model of optimality and select the plan that is most optimal. We formulate the plan optimization in the context of creating multiple bilingual dictionaries from a set of language of interest as a constraint optimization problem (COP). Formally, a constraint satisfaction problem is defined as a triple <X,D,C>, where $X = \{X_1, ..., X_n\}$ is a set of variables, $D = \{D_1, ..., D_n\}$ is a set of the respective domains of values, and $C = \{C_1, ..., C_m\}$ is a set of constraints [44]. A detailed definition of the variables, domains and set of constraints will be discussed in our future work.

3.4 Conclusion

In this chapter, we have discussed several bilingual dictionary creation methods and finally conclude that the pivot-based induction techniques: the inverse consultation method, the one-to-one mapping approach, the constraint-based induction method, and the generalized constraint-based induction method utilizing cognate recognition technique, are the most suitable and feasible methods to enrich the low-resource languages since bilingual dictionaries are the only language resource required as input. We formalize a comprehensive bilingual dictionaries creation planning that utilizes those techniques and a manual creation as constraint optimization problem with an objective to reduce total cost. A stochastic nature of the pivot-based bilingual lexicon induction is best handled by a Markov Decision Process (MDP), a well-known technique to solve problems containing uncertainty. Therefore, we will implement the plan optimization to bilingual dictionaries creation as a directed acyclic graph with MDP in our future work.

Acknowledgements This research was supported by Universitas Islam Riau (UIR) and Universiti Teknologi PETRONAS (UTP) Joint Research Program. This research was partially supported by a Grant-in-Aid for Scientific Research (A) (17H00759, 2017–2020) and a Grant-in- Aid for Young Scientists (A) (17H04706, 2017–2020) from Japan Society for the Promotion of Science (JSPS).

References

1. L. Campbell, Historical linguistics: the state of the art, in *Linguistics Today—Facing a Greater Challenge* (John Benjamins Publishing, 2004)
2. W.P. Lehmann, *Historical linguistics: an introduction* (Routledge, 2013)
3. L. Campbell, W.J. Poser, *Language Classification: History and Method* (Cambridge University Press, Cambridge, 2008)
4. M. Swadesh, Salish internal relationships. Int. J. Am. Linguist. **16**(4), 157–167 (1950)
5. M. Swadesh, Lexico-statistic dating of prehistoric ethnic contacts: with special reference to North American Indians and Eskimos. Proc. Am. Philos. Soc. **96**(4), 452–463 (1952)
6. M. Swadesh, *The Origin and Diversification of Language* (Routledge, Abingdon, 2017)
7. M. Swadesh, Towards greater accuracy in lexicostatistic dating. Int. J. Am. Linguist. **21**(2), 121–137 (1955)
8. V.I. Levenshtein, Binary codes capable of correcting deletions, insertions, and reversals. Sov. Phys. Dokl. **10**(8), 707–710 (1966)
9. B. Kessler, Computational dialectology in Irish gaelic, in *Proceedings of the Seventh Conference on European Chapter of the Association for Computational Linguistics* (1995), pp. 60–66
10. W.J. Heeringa, Measuring dialect pronunciation differences using Levenshtein distance, Doctoral dissertation, University of Groningen (2004)
11. C. Tang, V.J. Van Heuven, Predicting mutual intelligibility of Chinese dialects from multiple objective linguistic distance measures. Linguistics **53**(2), 285–312 (2015)
12. F. Petroni, M. Serva, Language distance and tree reconstruction. J. Stat. Mech Theor. Exp. **2008**(08), P08012 (2008)
13. S. Wichmann, E.W. Holman, D. Bakker, C.H. Brown, Evaluating linguistic distance measures. Phys. A Stat. Mech. Appl. **389**(17), 3632–3639 (2010)
14. E.W. Holman et al., Automated dating of the world's language families based on lexical similarity. Curr. Anthropol. **52**(6), 841–875 (2011)
15. M. Mundhenk, J. Goldsmith, C. Lusena, E. Allender, Complexity of finite-horizon Markov decision process problems. J. ACM **47**(4), 681–720 (2000)
16. R. Rapp, Identifying word translations in non-parallel texts, in *Proceedings of the 33rd Annual Meeting on Association for Computational Linguistics*, (1995), pp. 320–322

17. P. Fung, A statistical view on bilingual lexicon extraction: from parallel corpora to non-parallel corpora, in *Machine Translation and the Information Soup* (Springer, 1998), pp. 1–17
18. A. Irvine, A comprehensive analysis of bilingual lexicon induction. Comput. Linguist. **70**(8), 4943 (2017)
19. C. Schafer, D. Yarowsky, Inducing translation lexicons via diverse similarity measures and bridge languages, in *Proceedings 6th Conference Natural Language Learning*, vol. 20 (2002), pp. 1–7
20. A. Klementiev, A. Irvine, C. Callison-Burch, D. Yarowsky, Toward statistical machine translation without parallel corpora, no. 1995 (2012), pp. 130–140
21. R. Tanaka, Y. Murakami, T. Ishida, Context-based approach for pivot translation services, in *IJCAI International Joint Conference on Artificial Intelligence* (2009), pp. 1555–1561
22. T. Ishida, Y. Murakami, D. Lin, T. Nakaguchi, M. Otani, Language service infrastructure on the web: the language grid. Computer (Long. Beach. Calif) **51**(6), 72–81 (2018)
23. K. Tanaka, K. Umemura, Construction of a bilingual dictionary intermediated by a third language, in *Proceedings of the 15th Conference on Computational Linguistics*, vol. 1, 297–303 (1994)
24. X. Saralegi, I. Manterola, I. San Vicente, I.S. Vicente, Analyzing methods for improving precision of pivot based bilingual dictionaries, in *Proceedings of 2011 Conference Empirical Methods Natural Language Process* (2011), pp. 846–856
25. F. Bond, R.B. Sulong, T. Yamazaki, K. Ogura, *Design and Construction of a machine-tractable Japanese-Malay Dictionary* (1998)
26. F. Bond, K. Ogura, Combining linguistic resources to create a machine-tractable Japanese-Malay dictionary. Lang. Resour. Eval. **2**(2), 127–136 (2008)
27. V. István, Y. Shoichi, Bilingual dictionary generation for low-resourced language pairs, ... in *Methods Natural Language Processing*, vol. 2, (vol. 24) no. 4 (2009), pp. 862–870
28. D. Shezaf, A. Rappoport, Bilingual lexicon generation using non-aligned signatures, in *Proceedings of 48th Annual Meeting Association Computing Linguistic*, no. July (2010) pp. 98–107
29. J. Sjöbergh, Creating a free digital Japanese-Swedish lexicon, in *Proceedings of PACLING* (2005), pp. 296–300
30. Mausam et al., Compiling a massive, multilingual dictionary via probabilistic inference, in *Proceedings of the Joint Conference of the 47th Annual Meeting of the ACL and the 4th International Joint Conference on Natural Language Processing of the AFNLP*, Volume 1-Volume 1 (2009), pp. 262–270
31. T. Tsunakawa, N. Okazaki, J. Tsujii, Building bilingual lexicons using lexical translation probabilities via pivot languages, in *Proceedings of the 6th International Conference on Language Resources and Evaluation, LREC 2008* (2008), pp. 1664–1667
32. M. Wushouer, D. Lin, T. Ishida, K. Hirayama, A constraint approach to pivot-based bilingual dictionary induction. ACM Trans. Asian Low-Resour. Lang. Inf. Process. **5**(1), 1–26 (2015)
33. C. Ansótegui, M.L. Bonet, J. Levy, Solving (weighted) partial MaxSAT through satisfiability testing, in *Theory and Applications of Satisfiability Testing-SAT 2009*, ed. by O. Kullmann (Springer, 2009), pp. 427–440
34. A.H. Nasution, Y. Murakami, T. Ishida, Constraint-based bilingual lexicon induction for closely related languages, in *Proceedings of the 10th International Conference on Language Resources and Evaluation, LREC 2016*, vol. 16, no. 1955 (2016)
35. A.H. Nasution, Y. Murakami, T. Ishida, A generalized constraint approach to bilingual dictionary induction for low-resource language families. ACM Trans. Asian Low-Resour. Lang. Inf. Process. **17**(2), 1–29 (2017)
36. G.S. Mann, D. Yarowsky, Multipath translation lexicon induction via bridge languages, in *Proceedings of the Second Meeting of the North American Chapter of the Association for Computational Linguistics on Language technologies* (2001), pp. 1–8
37. R. van Bezooijen, C. Gooskens, R. Van Bezooijen, C. Gooskens, How easy is it for speakers of Dutch to understand Frisian and Afrikaans, and why? Linguist. Neth. **22**(1), 13–24 (2005)

38. C. Gooskens, Linguistic and extra-linguistic predictors of inter-Scandinavian intelligibility. Linguist. Neth. **23**(1), 101–113 (2006)
39. P. Koehn, K. Knight, Estimating word translation probabilities from unrelated monolingual corpora using the EM algorithm, in *AAAI/IAAI* (2000), pp. 711–715
40. I.D. Melamed, Automatic evaluation and uniform filter cascades for inducing N-best translation lexicons, in *Proceedings of Third Work. Very Large Corpora* (1995) p. 15
41. P. Danielsson, K. Muehlenbock, Small but efficient: the misconception of high-frequency words in Scandinavian translation, in *Envisioning Machine Translation in the Information Future* (Springer, 2000), pp. 158–168
42. D. Inkpen, O. Frunza, G. Kondrak, Automatic identification of cognates and false friends in French and English, in *Proceedings of the International Conference Recent Advances in Natural Language Processing* (2005) pp. 251–257
43. A.H. Nasution, Y. Murakami, T. Ishida, Plan optimization for creating bilingual dictionaries of low-resource languages, in *Proceedings of International Conference on Culture and Computing, IEEE* (2017), pp. 35–41
44. S.J. Russell, P. Norvig, *Artificial Intelligence: A Modern Approach* (Pearson Education Limited, 2016)

Chapter 4
Biomass Activated Carbon from Oil Palm Shell as Potential Material to Control Filtration Loss in Water-Based Drilling Fluid

Mursyidah Umar, Arif Rahmadani Amru, Nur Hadziqoh Muhammad Amin, Hasnah M. Zaid, and Beh Hoe Guan

Abbreviations

Å	Angstrong
°C	°Celcius
cP	centiPoise
lb	Pound
Ft	Feet
ml	Milliliter
mm	Millimeter

Activated carbon (AC) is a material that has powerful adsorptive properties. It has surface area excess 1000 m^2/g and porous structure. The properties of AC allow it to be a potential material to adsorb liquid and gas phase. This study is to investigate activated carbon from palm oil shell as a solution filtration loss in water based drilling fluid. Three general processes to produce activated carbon are (1) dehydration of water by heated in oven for 1 h at 100 °C. (2) carbonization by heated in oven for 1 h at 300 °C and (3) activation by heated in furnace for 1 h at 1000 °C. Scanning Electron Microscope (SEM) was used to determine the surface morphology of the activated carbon products and to find out if the carbon was active. Sample of water-based drilling fluid was prepared. Standard API filtration test was carried out to discover

M. Umar (✉) · A. R. Amru · N. H. M. Amin
Faculty of Engineering, Universitas Islam Riau, Jl. Kaharuddin Nasution No. 113 Perhentian, Marpoyan, Pekanbaru, Riau, Indonesia
e-mail: mursyidahumar@eng.uir.ac.id

H. M. Zaid · B. H. Guan
Fundamental and Applied Sciences Department, Universiti Teknologi PETRONAS, Seri Iskandar, Malaysia

© The Author(s), under exclusive license to Springer Nature Singapore Pte Ltd. 2020
S. A. A. Karim et al. (eds.), *Optimization Based Model Using Fuzzy and Other Statistical Techniques Towards Environmental Sustainability*,
https://doi.org/10.1007/978-981-15-2655-8_4

the effect of oil palm shell activated carbon on filtration loss drilling mud. The results show that filtration loss decreased from 21.3 ml with non-activated carbon to 15 ml with activated carbon. SEM shows that activated carbon has good pores and has good adsorption properties. These results indicate that activated carbon is a potential material to control filtration loss in the water-based drilling fluid.

4.1 Introduction

Drilling fluid which is also called drilling mud is an important process in the petroleum industry. Drilling fluid can affect the optimum drilling operations. The important thing in the drilling operations is the selection of fluid that will lower the cost and minimize the amount of lost time. There are several types of fluid in the drilling process, but the most commonly used is mud which is a mixture of water and clay. The water-based mud is more easily mixed with other additive materials; and it has economic value. The mud acts as a circulating fluid which functions as cutting from the well to the surface, maintains wellbore stability, controls formation pressure, cools and lubricates the drilling bits [1].

During the drilling process, keeping wells sustainability is very important. But the petroleum industry faces some of the challenges in drilling fluid process. Clay is a mixture of solid and liquid; when the clay is interacting with the formation which is porous medium, there will be pressure from the clay against the drill wall, it can cause fluid loss due to permeation into formation. The fluid that enters the formation is called filtrate. A way to reduce filtration loss during drilling process is to use additives [2]. Some materials can be used to prevent or reduce filtration loss used such as gilsonite, different nanoparticles such as bentonite, calcium carbonate [3], boehmite, XC polymer, nano metal oxide, nano zinc oxide, nano-iron oxide (III), nano silica, carbon nano structure [4].

In the recent study, the new issues of this field are multifunction of drilling fluid additives, novel environment-friendly drilling fluid system, and special materials for reservoir. One of the potential materials as additive in drilling fluid is activated carbon. Activated carbon is an environmentally friendly adsorbent that is widely used in industrial companies. Activated carbon can be found from various sources of carbonaceous materials such as coconut shell, sawdust, agricultural activities waste [5], plant shells, asphalt, and the rest of the polymer. Activated carbon has physical characteristic such as pore size distribution and pore structure with a surface area around $1000 \, m^2/g$. The quality of activated carbon is dependent on the raw materials and the activation process of the activated carbon. At present, activated carbon has been used in the development of non-destructive drilling fluids. The results show that activated carbon has good performance compared the calcium carbonate. It can reduce the filtrate value better than calcium carbonate. Activated carbon is not toxic material that has good adsorptive properties. Activated carbon compared to other material as an additive in Table 4.1 [6].

Table 4.1 Activated carbon properties

Material	Advantages	Weakness
Activated carbon	Inertness	The performance of activated carbon depends on raw materials
	Stability	The adsorption capacity of activated carbon is finite
	Large surface area	Activated carbon is not affective to adsorb minerals, salt and inorganic compounds
	Resistance	
	Low cost	

The objectives of the current studies are:

(1) To produce activated carbon from palm oil shell
(2) To analyse the effect of activated carbon to control filtration loss in water based drilling fluid.

4.2   Drilling Fluid

The drilling fluid which is also called drilling mud is an important part of the oil and gas exploration process. The drilling process is the most expensive part of the oil and gas industry. 25% of the total oilfield exploitation cost is in exploration and well drilling development. In the 90s, United State costs in drilling operations are about $10.9 billion with the total cost of the petroleum industry in exploration and production is $45.2 billion [7]. Drilling fluid is used in several functions simultaneously. They are intended to carry out drill cuttings, cool and lubricate the bit, minimize formation damage, and control formation pressures.

Drilling fluid has four types of phases:

1. The liquid phase can be either oil or water. And the water used can be divided into two, namely fresh water and salt water. This type of phase uses 75% of the water in the drilling fluid.
2. Reactive solids, these solids can react with their surroundings to form colloidal. In this case clay such as bentonite sucks (absorbs) fresh water and forms mud.
3. Solid inert (solids that cannot react), usually additive barite ($BaSO_2$) which is used to increase the density of fluid.
4. Chemical phase, chemicals are important parts that are used to control the properties of mud. Many chemicals are used commonly to reduce viscosity, reduce filtrate loss and control the colloidal phase (called the surface active agent).

Drilling fluid can be classified into three types, water-based mud, oil-based, and air or gas-based mud

a. Water-based mud

The composition of this sludge consists of fresh water or salt water, clay, and chemical additives. This composition is determined according to the conditions in the well borehole. Water-based mud is a type of mud that can form mud cake and is useful for keeping the drill holes from collapsing.

Drilling operations are improved by applying drilling mud There are three main types of drilling mud, mostly used in drilling operations, water-based mud, young oil based, and synthetic based mud. The choice of drilling mud depends on the behavior of the formation to be drilled. Water-based mud has more advantage compared oil-based mud and gas-based mud. It has high yield point value and decreased the loss of formation press. So that it can improve the stability of the borehole. The main constituent of water-based mud is water, thus all other constituents such as bentonite and barite are considered additives. Bentonite is often considered an additive drilling mud that is important because it provides viscosity and as filtration loss control.

b. Oil-based mud

This type of sludge is more expensive, but it can reduce the occurrence of corrosion processes that can cause fatal damage to the drill string circuit.

c. Air or gas-based mud

The advantage of this type of sludge is mainly that it can produce higher drilling rate. Because the tools commonly used are compressors which require a small number of equipment and small workspace only.

4.3 Formation Damages Influenced by Drilling Fluids

The petroleum industry faces some of the challenges in drilling fluid process. The challenges of drilling fluid process are shallow water flow problem, poorly consolidated formation, lost circulation, increase in torque and drag, differential pipe sticking, bit balling in gumbo shale, blowout situation in subsurface hydrate-bearing zones, emission of deadly acidic gas, high pressure and high-temperature wells.

Drilling fluid properties such as viscosity, permeability, density, plastic viscosity, and yield points play an important role in designing efficient and optimized drilling operations, hence it is important to ensure that drilling fluids have the right rheological properties. Controlling the flow characteristics and rate of filtration loss in drilling operations is an important aspect of drilling technology. This is because the occurrence of rheological changes in drilling mud has many effects on the level of efficiency used by mud to carry out its main functions

There are several types of fluid in the drilling process. The most commonly used is the mixture of water and clay. Water-based mud is more easily mixed with other additive materials and has economic value. Clay is a mixture of solid and liquid when the clay is interacting with borehole wall which is porous medium, there will be pressure from the clay against the drill wall, it can affect fluid loss due to permeation into the porous medium.

4.4 Filtration Loss

When drilling mud passes through a porous rock formation, the formation acts as a filter that allows small fluids and solids to pass through. The fluid that is lost or escapes into the rock is called the filtrate while the solid-solid layer deposited on the rock surface is called mad cake. If the filtration and mud cake is not properly considered, it can cause various problems. A thin Mud Cake will be a good cushion between the drilling pipe and the borehole surface. The thick mud cake can clamp the drilling pipe so that it is difficult to be lifted and rotated. Meanwhile, if too much filtrate enters the formation. It can cause formation damage.

For many reasons, the petroleum industry in the last 20 years has spent a lot of money and energy to determine the volume of filtrate drilling mud that enters rocks around the borehole and many researches put efforts to reduce the volume of this filtrate. Some reasons for the effort to determine the filtrate volume of drilling mud are as follows.

1. If the filtrate damages the permeability of the oil sand, the damage produced on the productivity of the oil well will depends on the distance of the filtrate invading the oil sand and the reduction in the volume of filtrate can increase the productivity of the well.
2. Filtrates that penetrate the shale can cause the shale to expand into the wellbore. If this problem cannot be controlled, it can cause the drill pipe to be pinched. Decreasing filtrate volume can reduce drilling problems.
3. The resistivity electric log curve changes because of the invasion of the drilling mud filtrate, the change depends on the depth of the invading filtrate. Knowing about this depth is needed before the resistivity log can be interpreted accurately.

There are two types of filtration, namely static filtration, dynamic filtration and understanding the mechanism of each type of filtration and its practical implications can cause a significant reduction in drilling mud filtrate without reducing the function of drilling mud.

Static filtration from drilling mud has long been recognized as an important parameter for drilling operations. Because standard laboratory testing procedures only consider static conditions, the filtration and mud cake properties in continuous circulation and dynamic filtration are very important for applying conditions below the actual surface.

Static filtration occurs when pumping mud drilling is disrupted. The interference creates a difference between hydrostatic pressure in the wellbore and in the reservoir, and from that problem, static filtration occurs. The level of static filtration is controlled by continuous mud cake thickening. On the other hand, dynamic filtration occurs when drilling mud is pumped through a well. In this process, the thickness of the mud cake is determined by the dynamic balance of two factors, namely the amount of deposited solid particles and the rate of erosion caused by shear stress produced by fluid flow in the wellbore.

According to fluid loss control in drilling mud can be used to:

a. Maintain the integrity of the wellbore hole
b. Reduce the fluid loss that occurs in productive formations

Some mud additives can be used to control filtration loss. Generally, these additives are used together with bentonite, while a small portion can be used separately in each clay content in the mud.

4.5   Activated Carbon from Oil Palm Shell

Palm oil is a natural source that is easy to obtain, cheap, non-toxic and not explosive. In recent years, researchers have succeeded in growing carbon nanotubes (cNTs) graphene on nickel substrate sourced from oil palm. In this research, the raw material of activated carbon using waste oil palm shell. The properties of activated carbon are:

1. Chemical properties
    Oxygen on the surface of activated carbon usually comes from the raw material itself or can occur when the activation process with steam (H_2O) or air is carried out. Generally, activated carbon contains mineral components. This component can become more concentrated when the carbon activation process is carried out. Chemicals commonly used in the activation process often cause changes in the chemical properties of the carbon produced so that the results are not optimal.
2. Physical properties
    Activated carbon has several characteristic properties such as solids that are black, odorless, insoluble in water, and acidic. Activated carbon cannot be damaged by the influence of temperature or the addition of pH during the activation process.
3. Pore structure
    The pore size of activated carbon depends on the activation temperature of the raw material that will become activated carbon. Pore size can vary, where the size can be divided into three categories:

    a. Macropores that have diameter sizes greater than 250 Å
    b. Mesopores that have diameter sizes ranging from 50 to 250 Å
    c. Micropores that have diameter sizes smaller than 50 Å

4. Adsorption
    Activated carbon adsorption is an event that accumulates or concentrates the interface in two phases. If the two phases interact, a new phase will be formed that is different from the previous phase. This event is caused by the attraction of the molecules of the two phases. This attraction is commonly known as the Van der Wall style. Under certain conditions atoms, the ions or molecules in the interface can experience a force imbalance so the molecules will continue to pull until the force balance is reached.

There are several factors that can affect the ability of adsorption on activated carbon including the nature of the components it absorbs, the nature of the solution and its contact system. Some other literature describes the absorption of activated carbon as physical absorption. This is because it has a uniform pore distribution and has a wide surface.

4.6 Material and Method

The raw material is palm oil shell from PT. Tunas Baru Lampung, Kecamatan Beringin, Kabupaten Pelalawan, Riau, Indonesia as shown in Fig. 4.1.

This research uses physical method to produce activated carbon. General process for producing activated carbon is shown in Fig. 4.2.

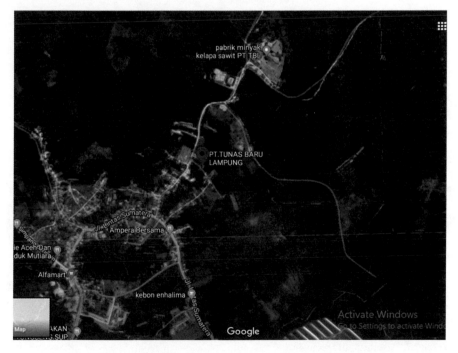

Fig. 4.1 Map of PT Tunas Baru Lampung

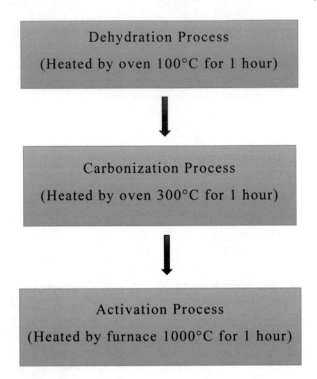

Fig. 4.2 General process to produce activated carbon

4.7 Result and Discussion

Rheological test and filtration loss tests were carried in drilling fluid to compare nonactivated carbon and activated carbon as an additive in the drilling process. The rheological test consists of plastic viscosity (μ_p), yield point (Y_p), and gel strength (GS) as shown in Table 4.2.

Based on API standard Spec 13A shown in Table 4.3, the maximum value of plastic viscosity and yield point is 3. Additive activated carbon is still in standard API Spec 13A [8] while additive non-activated carbon is not.

Table 4.2 The result of rheological test of drilling fluid using additive non-activated carbon and activated carbon

No.	Rheological test	Unit	Non-activated carbon	Activated carbon
1	Plastic viscosity (μ_p)	cP	4	2
2	Yield point (Y_p)	lb/100ft^2	1	2
3	Gel strength (GS)	lb/100ft^2	1	0.66

Table 4.3 Bentonite physical specifications

Requirement	Standard
Suspension properties	
Yield point/plastic viscosity ratio	Maximum 3
Filtrate volume	Maximum 15.0 ml

Table 4.4 Volume filtrate and mud cake of drilling fluid using additive non-activated carbon and activated carbon

No.	Sample	Volume filtrate (ml)	Mud cake (mm)
1	Non activated carbon	21.3	1.95
2	Activated carbon	15	1.33

Volume filtrate of drilling fluid using additive non activated carbon is 21.3 ml and using activated carbon is 15 ml as shown in Table 4.4. This value is under API standard spec 13A. so that additive activated carbon is a potential material to control filtration loss in the drilling process.

The difference of non-activated carbon compared to activated carbon is the non-activated carbon has a surface that is still covered by impurity deposits and has no pores, while the surface of activated carbon has been released from impurity deposits and has open pores so that the adsorption ability of activated carbon is higher than non-activated carbon. The surface image of material shown in Fig. 4.3.

4.8 Conclusion

Activated carbon from palm oil shell was produced. The result of the rheological test of drilling fluid with additive activated carbon is still in standard API Spec 13A. Filtration test shows that fluid loss decreases from 21.3 to 15 ml using additive activated carbon compared to additive non activated carbon. These results indicate that activated carbon from palm oil shell is a potential additive to control filtration loss in the drilling process.

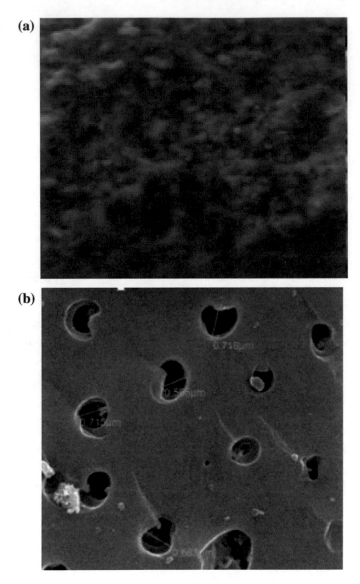

Fig. 4.3 SEM images of **a** non-activated material, **b** activated material

Acknowledgements The authors thank International Collaboratives Research Funding (ICRF) Universitas Islam Riau and Universiti Teknologi Petronas for financial support offered through the project No. 437/Kontrak/LPPM-UIR-9-2018.

References

1. A.A. Sulaimon, B.J. Adeyemi, M. Rahimi, Performance enhancement of selected vegetable oil as base fluid for drilling HPHT formation. J. Pet. Sci. Eng. **152**, 49–59 (2017)
2. P.P. Aydar, M.A. Hmadi, Characteristics of water-based drilling mud containing Gilsonite with Boehmite nanoparticles. Bull. la Société R. des Sci. Liège **86**, 248–258 (2017)
3. L. Li, X. Xu, J. Sun, X. Yuan, Y. Li, Vital role of nanomaterials in drilling fluid and reservoir protection applications (2012)
4. A.I. El-Diasty, A.M.S. Ragab, Applications of nanotechnology in the oil gas industry: latest trends worldwide and future challenges in Egypt, in *North Africa Technical Conference Exhibition* (2013),, pp. 1–13
5. S. McLean, Recent issues in assisted reproduction in the United Kingdom. Clin. Risk **9**(1), 18–22 (2003)
6. W.M.T.M. Reimerink, The use of activated carbon as catalyst and catalyst carrier in industria applications. Stud. Surf. Sci. Catal. V120A 751-69 **120**, 751–769 (1999)
7. M. Khodja, M. Khodja-Saber, J.P. Canselier, N. Cohaut, F. Bergaya, Drilling fluid technology : performances and environmental considerations, in *Products and Services, From R&D to Final Solut* (Igor Fuerstner, 2010), pp. 227–256
8. T. Vekemans, Table of contents. Double-Clicking Temple Bell Devot. Asp. Jainism online **6** (2014), pp. 126–143 (August 2010)

Chapter 5
Experimental Study of PV/T Air Solar Collector and Generation of Fuzzy Membership Functions

Mahmod Othman, Noran Nur Wahida Khalili, Mohd Nazari Abu Bakar, and Hamzah Sakidin

This study discussed the experimental study on the performance of a PV/T air solar collector. The operation of the solar collector is observed and studied under the change of parameters which are temperature, solar irradiance and air mass flow rate. The data collected is then used in the generation of fuzzy membership functions of the solar collector by using inductive reasoning. The generation scheme is based on minimum entropy method. The membership functions of each parameter are also tabulated and shown in graphs.

5.1 Introduction

Solar energy is one of the renewable energies which has the potential to meet a significant part of the world's energy demand. It is clean, reliable and environmentally friendly. The research on photovoltaic/thermal (PV/T) solar collector has begun as early as in mid 1970s [1]. Since then, the study has been developed from time to time. Researches have been done on many aspects of a collector, including the design, and the theoretical and experimental studies of the solar collector performances.

The technology of air-based PV/T innovation (PV plus thermal) is widely used and its known general operation efficiencies range from 20 to 40%. Since the efficiency of crystalline silicon cells ranging from 10 to 12%, hence the thermal part gives out the rest of the efficiency [2].

M. Othman (✉) · N. N. W. Khalili · H. Sakidin
Fundamental and Applied Sciences Department, Universiti Teknologi PETRONAS, Seri Iskandar, Malaysia
e-mail: mahmod.othman@utp.edu.my

M. N. Abu Bakar
Faculty of Applied Sciences, Universiti Teknologi MARA, Arau Campus, Arau, Perlis, Malaysia

© The Author(s), under exclusive license to Springer Nature Singapore Pte Ltd. 2020
S. A. A. Karim et al. (eds.), *Optimization Based Model Using Fuzzy and Other Statistical Techniques Towards Environmental Sustainability*,
https://doi.org/10.1007/978-981-15-2655-8_5

This study is to carry out experiments on the performance of a photo-voltaic/thermal model with air as a working fluid. The forced air will be produced by fan which speed can be controlled to keep the performance of the collector at optimum rate. The data collected is analysed and used in the generation of membership functions of the parameters involved in the experiment.

The main objective of the present studies is:

(a) To carry out experiment on the performance a PV/T air solar collector
(b) To generate membership functions of each parameter affecting the operation of solar collector

This chapter is organized as follows. Introduction and related literature review are written in Sect. 5.1. Section 5.2 discussed the experimental study on the performance of the PV/T solar collector while in Sect. 5.3, the generation of fuzzy membership function by using minimum entropy method is explained, and lastly Sect. 5.4 is dedicated for Results and Discussion.

5.2 Experiments on the Performance of PV/T Solar Collector

In this study, a design of a PV/T solar collector with air as a working fluid is presented (Fig. 5.1). The PV panel used is 50 W monocrystalline silicon of dimension

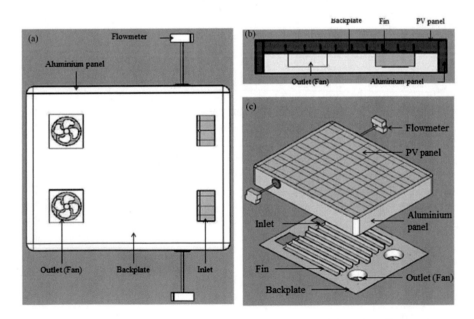

Fig. 5.1 **a** Back view, **b** cross-sectional view and **c** exploded view of the solar collector

69 cm × 54 cm × 3 cm. The PV panel comprises of solar cells encapsulated by a thin adhesive layer called ethylene-vinyl acetate (EVA), layered by tempered glass on top, and an absorber plate. The performance of the collector is improved by the use of set of fins as heat transfer enhancement parallel to the direction to the air flow. The back plate of the module, together with fins attached on it is made of aluminium. Aluminium is chosen as the material for the fins and back plate because of its ability to efficiently absorb and transfer heat from the solar cell, and also with consideration on the material density and economic viability [3]. There is an air gap for the air-flow between the absorber plate and the back plate. Air is pumped into the channel is created by two 12 W DC high flow rate fans attached parallel to the back plate. The speed of the fan is changed with different power resistors. The velocity of air is measured by a HK Instruments AVT air velocity transmitter. The temperatures of the PV module both at top and back surface, back plate and air were measured using LM35 precision centigrade temperature sensors. A Vernier pyranometer is used to measure the solar irradiance. All the temperature sensors, and the pyranometer were connected to an Arduino Mega 2560 for data logging via a PC.

5.2.1 Data Collection

The PV/T solar collector is set up for data collection in the compound of Universiti Teknologi PETRONAS. The set-up is based on the following conditions:

a. Mounting Location—PV modules can be mounted on ground, where power source is available
b. Shading—Photovoltaic arrays are adversely affected by shading. A well-designed PV system needs clear and unobstructed access to the sun's rays from about 9 a.m. to 3 p.m.
c. Orientation—PV modules are ideally oriented towards true south.
d. Tilt—The PV modules need to be installed according to the latitude angle of the location where it is placed. For the compound of UTP, the tilt angle is approximately 4.3590°.

The output parameters, current and voltage depends on two major input parameters, which are temperature and solar irradiance. The electrical current increases as the radiation increases. Likewise, the temperature of the cells also increases.

The performance of a solar collector is reflected by calculating its thermal and electrical efficiencies. The electrical efficiency, η_{ele} of the collector is modeled as a function of temperature based on [4] as follows:

$$\eta_{ele} = \frac{V_{OC} I_{SC} FF}{A_C G} \tag{5.1}$$

where fill factor, FF is

$$FF = (-0.132(T_0 + 75.36))/100 \tag{5.2}$$

The thermal efficiency, η_{th} of the collector by simplifying the equations by Abu Bakar [4] is:

$$\eta_{th} = \frac{\dot{m}C_f(T_0 - T_i)}{A_C G} \tag{5.3}$$

where mass flow rate, $\dot{m}$ and specific heat capacity of air, C_f is expressed as follows:

$$\dot{m} = v_{air}\rho_{air}C_L C_D \tag{5.4}$$

$$C_f = \left[1.0057 + 0.000066\left(\frac{T_p + T_{bp}}{2} - 300\right)\right]1000 \tag{5.5}$$

where density of air, ρ_{air} is

$$\rho_{air} = 1.774 - 0.00359\left(\frac{T_p + T_{bp}}{2} - 300\right) \tag{5.6}$$

Since there is difference in the nature of the electrical and thermal energy, the total overall thermal equivalent efficiency of a PV/T solar collector is calculated as:

$$\eta_{total} = \eta_{th} + \eta_{ele} \tag{5.7}$$

5.3 Minimum Entropy Method

Fuzzy logic control has been widely used in the system control and. The reason for using fuzzy logic in control applications stems from the idea of modeling uncertainties in the knowledge of a system's behavior through fuzzy sets and rules that are vaguely or ambiguously specified [5]. Fuzzy logic control requires fuzzy membership functions to be generated before fuzzification takes place. The intent of induction is to discover a law having objective validity and can be used universally where it begins with the particular and may be concluded with the general [6]. The induction is performed by using the entropy minimization principle, where the parameters corresponding to the output classes are most optimally clustered [7].

5.3.1 *Membership Functions Generation*

Membership function generation is based on the partitioning which draws a threshold line between two classes based on sample data. A threshold line is determined with an entropy minimization method, before segmentation process, first into two classes. Then by partitioning the first two classes one more time, three different classes are obtained. Therefore, by repeating the partitioning process, the data is classified into a number of classes, or fuzzy sets, depending on the shape of membership function used in each set.

The following is a review briefly explaining the threshold value calculation with induction principle for a two-class problem. Assuming that a threshold value for a sample is in the range between x_1 and x_2. Considering this sample alone, we write an entropy equation for the regions $[x_1, x]$ and $[x, x_2]$ regions. The first region is denoted p and the second region is q, as is shown in Fig. 5.2. By moving an imaginary threshold value x between x_1 and x_2, we calculate entropy for each value of x.

An entropy with each value of x in the region and is expressed by Christensen [1980] (as cited in [7]):

$$S(x) = p(x)S_p(x) + q(x)S_q(x) \tag{5.8}$$

where

$$S_p(x) = -[p_1(x) \ln p_1(x) + p_2(x) \ln p_2(x)] \tag{5.9}$$

$$S_q(x) = -[q_1(x) \ln q_1(x) + q_2(x) \ln q_2(x)] \tag{5.10}$$

where $p_k(x)$ and $q_k(x)$ are the conditional probabilities that the class k sample is in the region $[x_1, x_1 + x]$ and $[x_1 + x, x_2]$ respectively. $p(x)$ and $q(x)$ are the probabilities that all samples are in the region $[x_1, x_1 + x]$ and $[x_1 + x, x_2]$, respectively where $p(x) + q(x) = 1$.

A value of x that gives the minimum entropy is the optimum threshold value. The entropy estimates of $p_k(x)$, $q_k(x)$, $p(x)$ and $q(x)$ are as follows:

$$p_k(x) = \frac{n_k(x) + 1}{n(x) + 1} \tag{5.11}$$

Fig. 5.2 Illustration of threshold value idea

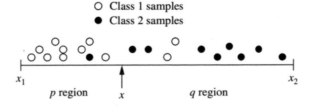

$$q_k(x) = \frac{N_k(x) + 1}{N(x) + 1} \tag{5.12}$$

$$p(x) = \frac{n(x)}{n} \tag{5.13}$$

$$q(x) = 1 - p(x) \tag{5.14}$$

where

$n_k(x)$ = number of class k samples located in $[x_1, x_1 + x]$
$n(x)$ = the total number of samples located in $[x_1, x_1 + x]$
$N_k(x)$ = number of class k samples located in $[x_1 + x, x_2]$
$N(x)$ = the total number of samples located in $[x_1 + x, x_2]$
n = total number of samples in $[x_1, x_2]$
l = a general length along the interval $[x_1, x_2]$

While moving x in the region $[x_1, x_2]$, the values of entropy for each position of x is calculated. The values of x that holds the minimum entropy is called the primary threshold (PRI) value. Then, the region $[x_1, x_2]$ is divided into two using the PRI value. The left side of the primary threshold is the negative side while the right side is the positive side. Then, a shape for the two membership functions are chosen, such as trapezoid, triangle or rectangle. The choice of shape is arbitrary. As the region $[x_1, x_2]$ is further subdivide, the choice of shape becomes less important, as long as the sets are overlapping.

Then in next sequence, the segmentation is repeated on each side of region to determine the secondary threshold values by applying the same procedure (Fig. 5.3a). The secondary threshold in the negative area is denoted as SEC1 and the other in the positive are is SEC2, therefore now there are three threshold lines in the sample space. The threshold SEC1 and SEC2 are the minimum entropy points that divide the respective areas into two classes. Then, three labels of PO (positive), ZE (zero) and NG (negative) are used for each of the classes, and the three threshold values (PRI, SEC1 and SEC2) are used as the toes of the three separate membership shapes as in Fig. 5.3b.

In fuzzy logic applications, odd number of membership functions are usually used to partition a region such as five or seven labels. Therefore, to develop seven partitions, tertiary threshold values are required in each of three classes. So, there are now four tertiary threshold values: TER1, TER2, TER3 and TER4. Two of the tertiary thresholds lie between primary and secondary thresholds and the other two is between secondary threshold and the ends of sample space such as shown in Fig. 5.3c. In this figure, the labels used are NB (negative big), NM (negative medium), NS (negative small), ZE (zero), PS (positive small), PM (positive medium) and PB (positive big).

Fig. 5.3 Repeated partitions and corresponding fuzzy sets label: **a** the first partition, **b** the second partition and **c** the third partition

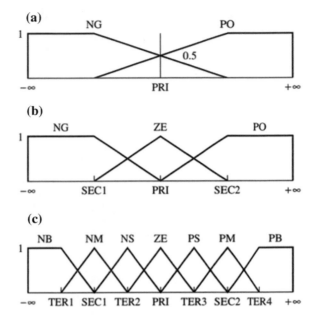

5.3.2   Fuzzification

The fuzzification is a coding process in which each numerical input of a linguistic variable is transformed in the membership function values of its attributes. First phase of fuzzy logic proceeding is to deliver input parameters for given fuzzy system based on which the output result will be calculated. Most variables existing in the real world are crisp or classical variables. One needs to convert those crisp variables (both input and output) to fuzzy variables, and then apply fuzzy inference to process those data to obtain the desired output. These parameters are fuzzified with the use of pre-defined input membership functions, which can have different shapes. The most common are triangular shape, however bell, trapezoidal, sinusoidal and exponential can be also used. The degree of membership function is determined by placing a chosen input variable on the horizontal axis, while vertical axis shows quantification of grade of membership of the input variable. The only condition a membership function must meet is that it must vary between zero and one. The value zero means that input variable is not a member of the fuzzy set, while the value one means that input variable is fully a member of the fuzzy set.

The triangular function $\mu(x)$ that is used to determine the membership function is defined by a lower limit a, an upper limit b, and a value m where $a < m < b$ (Fig. 5.4) as expressed in Eq. (5.15), while the trapezoidal function is in Eq. (5.16).

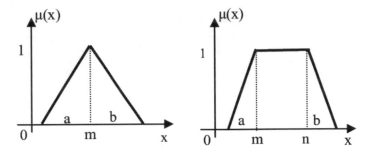

Fig. 5.4 Triangular and trapezoidal shape membership function

$$\mu(x) = \begin{cases} 0 & x < a \\ \frac{x-a}{m-a} & a < x \le m \\ \frac{b-x}{b-m} & m < x < b \\ 0 & x \ge b \end{cases} \qquad (5.15)$$

$$\mu_{Trap}(x) = \begin{cases} 0 & x < a \\ \frac{x-a}{m-a} & a \le x < m \\ 1 & m \le x < n \\ \frac{b-x}{b-n} & n \le x < b \\ 0 & x \ge b \end{cases} \qquad (5.16)$$

5.4   Results and Discussions

5.4.1   Experimental Results

The data collected during the experimental study is analyzed. The parameters that are considered are the temperature of the collector layers (temperature of PV panel T_p, temperature of backplate T_{bp}, temperature of fins T_{fin}, temperature of output air T_f and ambient temperature T_a), solar irradiance G and the air velocity v, meanwhile the output is the total efficiency, η_{total} of the solar collector.

The solar irradiance measured increases and decreases following a bell shape form while the temperature of each solar collector layers increases with time during the experiment (Figs. 5.5 and 5.6). It has also been observed that when solar irradiance increases, the current produced by the PV cells increases as shown in Fig. 5.7. In relation to this, as shown in Fig. 5.8, when current production increases, the electrical efficiency of the collector also increases. However, when solar irradiance and current increases, the temperature of the collector also increase, which leads to the drop in its thermal efficiency (Fig. 5.9). This is because solar cells are very sensitive towards

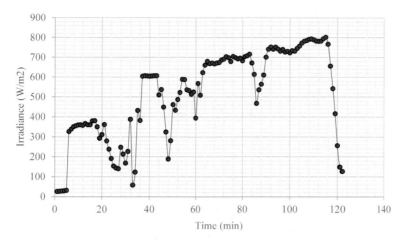

Fig. 5.5 Measured solar irradiance during experiment

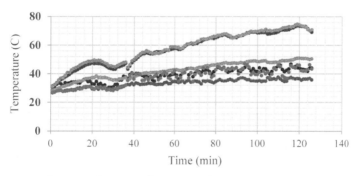

• Ave Top • Ave Bottom • Fin • Ave Inside • Ave Fan • Ambient Temp

Fig. 5.6 Temperature change of solar collector layers

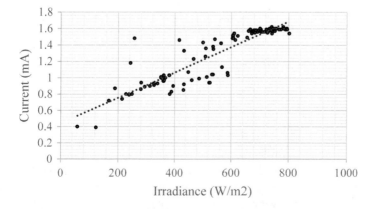

Fig. 5.7 Current change on the effect of irradiance

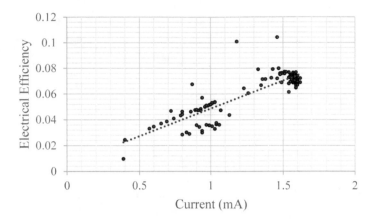

Fig. 5.8 Effect of current production on electrical efficiency of solar collector

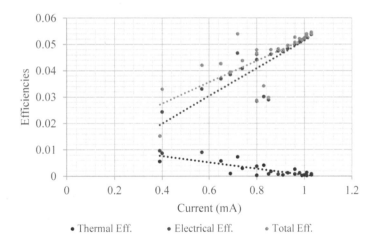

Fig. 5.9 Effect of current production on efficiencies of solar collector

temperature increase, where the its performance is reduced when there is temperature rise.

5.4.2 Numerical Example

The following is the sample calculation of the threshold values for solar irradiance parameter by using minimum entropy. Firstly, 78 solar irradiance values ranging from 133 to 788 W/m^2 were recorded and tabulated. All the solar irradiance data is arranged numerically, and the repeated values are removed. The data for solar

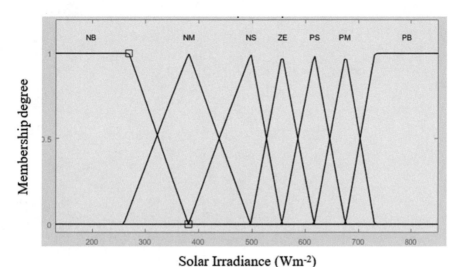

Fig. 5.10 Membership function for solar radiance

radiance that is less than 554 W/m^2 is clustered as Class 1, while the rest of the data is clustered into Class 2.

Then the values of x are selected as approximately the midvalue between any two adjacent values. Equation (5.1)–(5.7) are used to compute p_1, p_2, q_1, q_2, $p(x)$, $q(x)$, $S_p(x)$, $S_q(x)$ and S. The value of x that gives the minimum value of entropy, S is selected as the first threshold partition point PRI. The first partition point for solar irradiance is $x = 556\,\text{Wm}^{-2}$.

The same process is repeated for the partition of both negative and positive side for different values of x. The final results of the fuzzy partitions are written in Eqs. (5.17a)–(g) below and shown graphically in Fig. 5.10.

$$\mu_{NB}(x) = \begin{cases} 1 & x < 306.87 \\ \frac{307.47-x}{307.47-306.87} & 306.8 \leq x < 307.47 \\ 0 & x \geq 308.22 \end{cases} \tag{5.17a}$$

$$\mu_{NM}(x) = \begin{cases} 0 & x < 259.5 \\ \frac{x-259.5}{381.5-259.5} & 259.5 \leq x < 381.5 \\ \frac{497.5-x}{497.5-381.5} & 381.5 \leq x < 497.5 \\ 0 & x \geq 497.5 \end{cases} \tag{5.17b}$$

$$\mu_{NM}(x) = \begin{cases} 0 & x < 381.5 \\ \frac{x-381.5}{497.5-381.5} & 381.5 \leq x < 497.5 \\ \frac{556-x}{556-497.5} & 497.5 \leq x < 556 \\ 0 & x \geq 556 \end{cases} \tag{5.17c}$$

$$\mu_{NM}(x) = \begin{cases} 0 & x < 497.5 \\ \frac{x-497.5}{556-497.5} & 497.5 \le x < 556 \\ \frac{-x}{617-556} & 556 \le x < 617 \\ 0 & x \ge 617 \end{cases} \qquad (5.17d)$$

$$\mu_{NM}(x) = \begin{cases} 0 & x < 556 \\ \frac{x-556}{617-556} & 556 \le x < 617 \\ \frac{676-x}{676-617} & 617 \le x < 676 \\ 0 & x \ge 676 \end{cases} \qquad (5.17e)$$

$$\mu_{NM}(x) = \begin{cases} 0 & x < 617 \\ \frac{x-617}{676-617} & 617 \le x < 676 \\ \frac{730.5-x}{730.5-676} & 676 \le x < 730.5 \\ 0 & x \ge 730.5 \end{cases} \qquad (5.17f)$$

$$\mu_{NB}(x) = \begin{cases} 0 & x < 676 \\ \frac{730.5-x}{730.5-676} & 676 \le x < 730.5 \\ 1 & x \ge 730.5 \end{cases} \qquad (5.17g)$$

The calculations of minimum entropy for the other parameters (temperature, voltage and current) are not included to have a brief note.

The membership functions for temperature and current production is shown in Figs. 5.11 and 5.12. For air velocity, the speed of fan is divided into 4 levels, which are low, medium, high and fast as shown in Fig. 5.13.

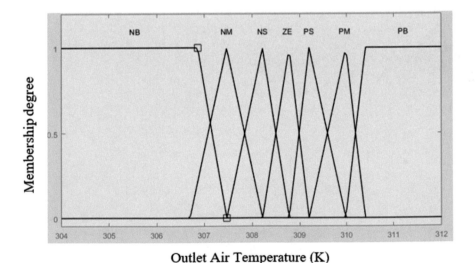

Fig. 5.11 Membership function for outlet air temperature

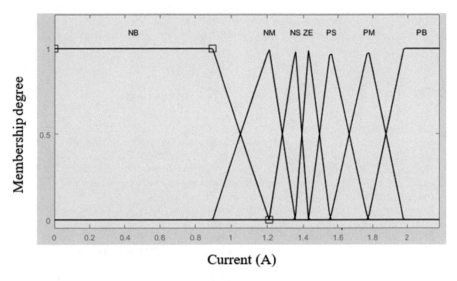

Current (A)

Fig. 5.12 Membership function for current production

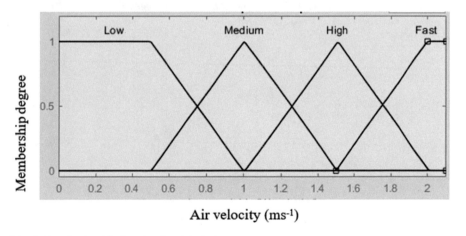

Air velocity (ms⁻¹)

Fig. 5.13 Membership function for air velocity

5.5 Summary

A photovoltaic/thermal solar collector has been set up and experiments have been carried out to assess its performance under changes of input parameters such as temperature and solar irradiance. It has been found out that the temperature of the layers of solar collector increases with solar radiation. In the same time, current produced by the PV panel also increases with irradiance. The electrical efficiency is found to be increasing with current, however the thermal efficiency drops when the temperature increases. The data collected is used to develop membership function of each parameter. By the minimum entropy method used, the membership functions for solar irradiance, temperature and current production were developed.

References

1. T.T. Chow, G. Pei, K.F. Fong, Z. Lin, A.L.S. Chan, J. Ji, Energy and exergy analysis of photovoltaic–thermal collector with and without glass cover. Appl. Energy **86**(3), 310–316 (2009)
2. M. Hazami, A. Riahi, F. Mehdaoui, O. Nouicer, A. Farhat, Energetic and exergetic performances analysis of a PV/T (photovoltaic thermal) solar system tested and simulated under to Tunisian (North Africa) climatic conditions. Energy **107**, 78–94 (2016)
3. R. Kumar, M.A. Rosen, Performance evaluation of a double pass PV/T solar air heater with and without fins. Appl. Therm. Eng. **31**(8–9), 1402–1410 (2011)
4. M.N. Abu Bakar, M. Othman, M.H. Din, N.A. Manaf, H. Jarimi, Design concept and mathematical model of a bi-fluid photovoltaic/thermal (PV/T) solar collector. Renew. Energy **67**, 153–164 (2014)
5. R.M. Aguilar, V. Muñoz, Y. Callero, Control application using fuzzy logic: design of a fuzzy temperature controller, in *Fuzzy Inference System-Theory and Applications* (InTech, 2012)
6. C.J. Kim, B.D. Russell, Automatic generation of membership function and fuzzy rule using inductive reasoning, in *Third International Conference on Industrial Fuzzy Control and Intelligent Systems* (IEEE, 1993), pp. 93–96
7. T.J. Ross, *Fuzzy Logic with Engineering Applications*, vol. 2 (Wiley, New York, 2004)

Index

© The Author(s), under exclusive license to Springer Nature Singapore Pte Ltd. 2020   81
S. A. A. Karim et al. (eds.), *Optimization Based Model Using Fuzzy and Other Statistical Techniques Towards Environmental Sustainability*,
https://doi.org/10.1007/978-981-15-2655-8

Printed in the United States
By Bookmasters